CRC Series
in
The Biochemistry and Molecular Biology
of the
Cell Nucleus

Editor-in-Chief

Lubomir S. Hnilica, Ph.D.
Professor of Biochemistry and Pathology
Department of Biochemistry
Vanderbilt University School
of Medicine
Nashville, Tennessee

**The Structure and
Biological
Function of Histones**
Author
Lubomir S. Hnilica, Ph.D.

**Chromosomal
Nonhistone Proteins**
Volume I: Biology
Volume II: Immunology
Volume III: Biochemistry
Volume IV: Structural
Associations
Editor
Lubomir S. Hnilica, Ph.D.

**Enzymes of Nucleic Acid
Synthesis and Modification**
Volume I: DNA Enzymes
Volume II: RNA Enzymes
Editor
Samson T. Jacob, Ph.D.
Professor
Department of Pharmacology
The Pennsylvania State University
The Milton S. Hershey Medical Center
Hershey, Pennsylvania

Chromosomal Nonhistone Proteins

Volume I
Biology

Editor

Lubomir S. Hnilica, Ph.D.

Profesor of Biochemistry and Pathology
Department of Biochemistry
Vanderbilt University School
of Medicine
Nashville, Tennessee

**CRC Series in The Biochemistry and Molecular Biology
of the Cell Nucleus**

CRC Press
Taylor & Francis Group
Boca Raton London New York

CRC Press is an imprint of the
Taylor & Francis Group, an **informa** business

First published 1983 by CRC Press
Taylor & Francis Group
6000 Broken Sound Parkway NW, Suite 300
Boca Raton, FL 33487-2742

Reissued 2018 by CRC Press

© 1983 by Taylor & Francis Group.
CRC Press is an imprint of Taylor & Francis Group, an Informa business

No claim to original U.S. Government works

A Library of Congress record exists under LC control number: 82017742

Publisher's Note
The publisher has gone to great lengths to ensure the quality of this reprint but points out that some imperfections in the original copies may be apparent.

Disclaimer
The publisher has made every effort to trace copyright holders and welcomes correspondence from those they have been unable to contact.

ISBN 13: 978-1-315-89157-6 (hbk)
ISBN 13: 978-1-351-07067-6 (ebk)

Visit the Taylor & Francis Web site at http://www.taylorandfrancis.com and the
CRC Press Web site at http://www.crcpress.com

INTRODUCTION

The first volume of the *Chromosomal Nonhistone Proteins* treatise presents a summary of the many attempts in the literature to correlate changes in chromosomal nonhistone proteins specificity and metabolism with transcriptional regulations in eukaryotic cells. Chapters on gene regulation, development, differentiation and their changes in response to hormonal stimulation or chemical carcinogenesis are presented together with the discussion of chromosomal nonhistone proteins which form meaningful associations with DNA. Skillful presentations by the individual authors will lead the reader through the many difficulties facing students of chromosomal proteins biology and expose the long and rocky road to the discovery of their roles and functions. *Ars longa, vita brevis est.*

THE EDITOR

Lubomir S. Hnilica is a professor of biochemistry and pathology at the Vanderbilt University School of Medicine, Nashville, Tennessee.

Dr. Hnilica received his degree in organic chemistry from Masaryk University, Brno, Czechoslovakia in 1952 and an advanced degree in biochemistry from the Czechoslovak Academy of Sciences. After postdoctoral training at the Czechoslovak National Cancer Institute in Bratislava, Dr. Hnilica continued his research as a WHO research fellow at the Chester Beatty Research Institute in London, England. After 2 years at Baylor College of Medicine in Houston, Texas, Dr. Hnilica joined the staff of the University of Texas System Cancer Center, M.D. Anderson Hospital and Tumor Institute in Houston where he rose to the rank of Professor of Biochemistry and Chief of the Section of Biochemical Regulatory Mechanisms. In 1975 Dr. Hnilica joined the Vanderbilt University School of Medicine where he is Professor of Biochemistry and Pathology as well as director of the A. B. Hancock, Jr. Memorial Laboratory of the Vanderbilt University Cancer Center. His present research concerns proteins of the cell nucleus, their biochemistry and immunology, the interactions of nuclear proteins with DNA, and changes in gene expression during chemical carcinogenesis.

CONTRIBUTORS

Michael Gronow, Ph.D.
Managing Director
Cambridge Life Sciences, P.L.C.
Cambridge, England

Ian R. Phillips
Department of Biochemistry
University College London
London, England

Mark A. Plumb
Department of Biochemistry and
  Molecular Biology
University of Florida
Gainesville, Florida

Warren N. Schmidt, Ph.D.
Research Instructor
Department of Biochemistry and the
  A. B. Hancock, Jr. Memorial
  Laboratory of the Vanderbilt
  University Cancer Center
Vanderbilt University School of
  Medicine
Nashville, Tennessee

J. Sanders Sevall, Ph.D.
Senior Scientist II
Department of Molecular Genetics
Wadley Institutes of Molecular
  Medicine
Dallas, Texas

Elizabeth Anne Shephard
Department of Biochemistry
University College London
London, England

Thomas C. Spelsberg, Ph.D.
Professor and Head
Section of Biochemistry
Mayo Graduate School of Medicine
Rochester, Minnesota

Gary S. Stein, Ph.D.
Professor of Biochemistry and
  Molecular Biology
Department of Biochemistry and
  Molecular Biology
University of Florida
College of Medicine
Gainesville, Florida

Janet L. Stein, Ph.D.
Associate Professor
Department of Immunology
  and Medical Microbiology
University of Florida
College of Medicine
Gainesville, Florida

TABLE OF CONTENTS

CHROMOSOMAL NONHISTONE PROTEINS

TABLE OF CONTENTS

Volume I

Chapter 1

NUCLEAR PROTEINS IN DIFFERENTIATION AND EMBRYOGENESIS

Warren Schmidt

TABLE OF CONTENTS

I. INTRODUCTION

The ordered progression of events which transform the pluripotent cells of the implanting blastocyst into the diversity of cell types characterisic of the complete organism can be justly referred to as one of the fundamental enigmas of biology. This series of processes, collectively referred to as cellular differentiation, is operationally defined as the progressive and selective manifestation of specialized functions and phenotypes particular to each cell type of a given organism. This definition is of course very inclusive and contains aspects of nearly all cellular processes, normal growth, and neoplasia. In this review the discussion will be restricted for the most part to development and embryogenesis of nonneoplastic cells.

It is important to remember that cellular differentiation ensues in general without a quantitative change in the DNA of the genome. The constancy of DNA among diploid cells of the same organism was demonstrated originally in the classical studies of Boivin et al.[1] and Mirsky and Ris[2] in the latter half of the 1940s. The concept has since been vindicated with a variety of studies including the elegant transplantation experiments of Gurdon[3,4] and further on a molecular level with the original DNA-DNA reassociation studies of McCarthy and Hoyer.[5] It is apparent that the genetic control of cellular differentiation cannot be explained easily through selective losses of nontranscribed portions of the genome. Moreover, only minor fractions of the genome in differentiated cells are actually represented in functional mRNA. Quite simply, only the expression of specific DNA segments is controlled during differentiation and the process likely depends upon selective gene activation, i.e., events occurring at the transcriptional and/or translational level.

It is generally believed that the model of genetic expression proposed by Jacob and Monod,[6] will hold for eukaryotic cell regulation, that is, that the genome may be repressed or derepressed by regulatory molecules acting on DNA and either inhibiting or allowing transcription. Although the exact nature of these regulatory molecules remains to be identified in eukaryotic cells, an increasing amount of evidence indicates the nuclear proteins to fulfill at least a part of this role. Thus, if differentiation involves changes in the genetic expression of cells, we might ask ourselves what type of alterations can be observed in the nuclear proteins in part responsible for and accompanying genetic expression during cellular differentiation?

Chromatin is composed chiefly of DNA, a lesser quantity of RNA, and in general two classes of nuclear protein, the histones and nonhistones. In this review attention is focused mainly on the nonhistones and the histones will be only examined briefly since an increasing body of evidence indicates these proteins to fulfill mainly a structural role in the nucleus. It is also important to keep in mind that chromatin is defined operationally by the method used in its preparation. A number of nuclear protein fractions, some likely important in cellular differentiation, may be either enriched or diminished depending on the method of chromatin isolation.

II. THE HISTONES

Of the two main categories of nuclear proteins the histones are by far the most well characterized.[7,8] They are small (mol wt about 20,000), metabolically stable, and highly basic polypeptides. Five principal classes have been detected throughout the plant and animal kingdoms, (H1, H2A, H2B, H3, and H4) and in a few instances an additional distinct species has been identified (e.g., H5 in avian erythrocytes).

One is impressed with the evolutionary conservation of the amino acid sequences of the histones, especially the arginine-rich fractions. Only two amino acid residues differ between histones H4 of pea bud or calf thymus.[9] On the other hand, H1 histones (lysine-rich), are an exception to this evolutionary conservation. These fractions are heterogeneous, with species and tissue specificity being observed for their amino acid sequences and immunological properties.[7,8,10,11] As a class however, the evolutionary stability of histones suggests that they would be poor choices as regulatory molecules in cellular differentiation, but that they fulfill a common but highly important structural role in chromatin — one that would tend to protect the proteins from mutational variability.

Although quantitative levels of histones vary during early embryogenesis and differentiation, the same protein species in all cell types are observed from the gastrula stage onward.[12-15] The tight coupling of histone and DNA synthesis in the division cycle of a typical eukaryotic cell is not apparent in all animal embryos. In early cleavage stages of *Xenopus* embryos, histone synthesis is rapid and exceeds DNA synthesis manifold. However, DNA synthesis increases at a greater rate during the early hours postfertilization such that synthesis of both DNA and histone is nearly equivalent by 5 to 6 hr. Thereafter, DNA synthesis increases creating a subsequent deficit of histone which is apparently equivalented with the amount of DNA by a pool of histone made previously during oogenesis.[16,17] It is unclear at the present what aspects of the amphibian pattern of histone synthesis applies to other groups of animals, however the uncoupling of DNA and histone synthesis during early embryogenesis in *Xenopus* poses interesting questions for chromatin assembly, replication, and cell division.

Some of the earliest studies performed on transcriptional regulation in eukaryotic cells indicated that histones are potent repressor molecules of DNA-dependent RNA synthesis,[18-20] but that the repression is generally nonspecific.[21-23] The limited heterogeneity of histones makes it unlikely that the molecules by themselves can recognize specific gene loci. Possibilities exist, however, that posttranslational modifications, i.e., phosphorylation, acetylation, and methylation may confer to histones greater specificity on gene repression through alterations of histone-histone, histone-DNA, or histone-nonhistone interactions. These modifications have been associated with changes in the transcriptional and structural properties of chromatin and have been observed during cellular development, differentiation, transformation with viruses, and hormonal stimulation.[7,8,14] Posttranslational modifications of all nuclear proteins have the potential for producing local alterations in chromatin structure resulting in activation of specific genes particular to a differentiated state and are reviewed in other chapters of this book.

Experimental interest on the function of histones has recently been directed at elucidation of their role in chromatin structure. Based on electron microscopy and biochemical evidence from a number of laboratories, histones appear to be packaged into an octamer containing two each of the four core histones (H2A, H2B, H3, H4) associated with approximatley 145 base pairs of DNA.[24,25] These "n" bodies or nucleosomes are arranged like "beads on a string", with the internucleosomal spacing region consisting of nuclease-sensitive DNA and also containing one molecule of H1 histones. Of particular interest is that histones do not appear to associate with defined sequences of DNA. Isolated nucleosomal DNA has the same apparent sequence complexity as that of total cellular DNA and in vitro transcriptional studies indicate that nucleosomal DNA has the same frequency of sequences coding for mRNAs.[26] These data suggest that nucleosome association is at random but equally spaced sites along the DNA and that some transcriptional sites are found among the histone-nucleosome complexes.

In summary, the emerging role of histones in eukaryotic cells is one chiefly of structure, however further studies are needed to clarify the relationships of nucleosome structure, transcriptional regulation, and the function of these properties in differential gene expression during cellular differentiation. Structure and function are two concepts that in chromatin research are often difficult to separate.

III. THE NONHISTONES

In contrast to the histones, nonhistone chromosomal proteins (NHCP) exhibit considerable heterogeneity. Whereas the eukaryotic cell nucleus can be shown to contain usually only five types of histones, the number of NHCP that can be detected is about 20 to several hundred depending upon the analytical system used and the metabolic and mitotic status of the chromatin source. The isolation, separation, and chemical properties of the NHCP have been the subject of several reviews[8,27-31] including other chapters in this book and will not be extensively dealt with here. It must be emphasized that NHCP have proven to be difficult to isolate and study because of their poor solubility, heterogeneity, and their tendency to aggregate with histones, themselves, and DNA.

The diversity and heterogeneity displayed by the NHCP however makes them especially attractive as candidates for a role as regulators of the eukaryotic genome during differentiation. It is well known that NHCP possess the ablity to stimulate or repress DNA-dependent RNA synthesis in vitro[20,32,33] which can be listed as a primary requisite for macromolecules to influence the programmed expression of different genes. Nevertheless, identification and characterization of the proteins responsible for transcriptional gene control or elucidation of the proteins' mechanism(s) of action have in general not been achieved.

NHCP have been implicated in a preponderant number of nuclear functions and potential gene regulatory processes. This author has arbitrarily divided the types of studies conducted with NHCP into four broad areas:

1. **Qualitative and quantitative changes** — Differences in the amount and composition of NHCP have been examined during in vivo or in vitro cell differentiation and maturation. In some instances the acquisition of a physiological function or morphological feature characteristic of the differentiated state has been correlated with the appearance or disappearance of various NHCP species.
2. **Tissue and cell specificity** — NHCP have been assessed from a number of sources both mature and embryonic with the overall aim to identify tissue and cell-specific proteins. NHCP demonstrating tissue specificity have the greatest potential for being involved in some aspect of the specific gene expression which accompanies differentiation.
3. **Control of specific mRNA synthesis** — Although all cellular differentiation is probably the result of activation or repression of specific genes, several systems (most notably those of the chick oviduct and avian erythrocyte), are particularly amenable to the analysis of specific mRNA synthesis. In these systems isolated fractions of NHCP have been examined for their ability to influence the activity of specific genes which are expressed transiently during differentiation.[34-39]
4. **Posttranslational modifications** — Like the histones, NHCP have been observed to undergo methylation, acetylation, phosphorylation, and other modifications in response to appropriate stimuli. These modifications have been correlated with a variety of nuclear regulatory mechanisms.[40-45]

The remainder of this review will be concerned primarily with the top two subheadings, i.e., with studies attempting to identify specific nonhistones during differentiation. For a thorough discussion of the latter two topics the reader is referred to the references listed as well as other chapters in this book.

A. Qualitative and Quantitative Variations in NHCP

Cellular differentiation and organogenesis are reflections of the altered expression of specific genes and various studies have investigated the qualitative and quantitative differences in NHCP during induced differentiation, normal embryogenesis, and after various other alterations in the biological states of developing cells. It must be recognized that while these studies suggest a regulatory role for NHCP in these processes the evidence is correlative and largely circumstantial.

The sea urchin has proven to be a valuable model for the examination of stage-related changes in NHCP during embryogenesis. This animal can be easily raised in the laboratory from fertilization through metamorphosis to adult.[46] During early embryonic development a progressive increase in the mass ratio of NHCP relative to DNA ensues such that by the pluteus stage this ratio is about one.[17,47-49] The increase in NHCP during embryogenesis has been confirmed with studies demonstrating a progressive enrichment of newly synthesized NHCP.[50] This new NHCP is likely responsible for a noticeably less compact structure for gastrula vs. blastula chromatin[50] and for increases observed in DNA-dependent RNA synthesis during the same developmental period.

In an analysis of the individual protein species present in chromatin at early embryonic periods, Cognetti et al.[51] observed qualitative differences in some NHCP components present at the blastula vs. the early pluteus stage as assessed with sodium dodecylsulfate (SDS) gel electrophoresis. These findings were somewhat confirmatory to the earlier results of Hnilica and Johnson[52] who measured changes in the amino acid composition of nuclear acidic and residual protein fractions during comparable developmental periods. In contrast however, Hill et al.[47] and Seale and Aronson[53] noted only quantitative differences in various NHCP species although slightly later time points in embryogenesis were examined. In addition, Gineitis et al.[48] reported only quantitative changes in NHCP during urchin embryogenesis using isolectric focusing in 5 M urea gels. In these studies embryos were induced experimentally to undergo animalization or vegetalization and these manipulations were found not to change dramatically the nature of NHCP species. On the other hand, Poccia and Hinegardner[49] identified both qualitative and quantitative differences in NHCP components which are present in the blastula compared to those present in the larval feeding stages after the pluteus stage. A definite compositional shift was noted during this transition period in that early embryogenesis was characterized by having NHCP species of relatively high molecular weight, (>80,000 daltons), while after the pluteus stage major bands were present in the midrange of molecular weights (40,000 to 80,000). Differences among the above studies may be ascribed to variations in the methods used for NHCP solubilization and separation from DNA as well as the resolving power of different gel electrophoresis systems. As we shall see in the ensuing discussions, these are major drawbacks of these types of studies, and prevent firm conclusions to be drawn as to the nature of the changes in NHCP during development.

Working with *Drosophila* embryos, Elgin and Hood[54] observed changes in NHCP in 2 to 6- vs. 6 to 18-hr embryo chromatin. The apparent acquisition of several new components occurred concomitantly with decreases in chromatin template activity as assessed with *Escherichia coli* RNA polymerase and an increase in chromatin melting temperatures.[54] The polytene chromosomes of *Drosophila* and some other insect species represent extremely useful systems for the correlation of nuclear events during embryogenesis and development. Stage-specific activation of a particular gene on a polytene chromosome is accompanied by "puffing" of restricted chromosomal regions and the DNA in the chromosomal band appears to unfold as RNA synthesis is initiated.[55] Puffing is very stage-specific and can be induced experimentally from a variety of stimuli such as heat-shock or the hormone ecdysone.[55] Of particular interest is that the gene activation involved in puffing is accompanied by a large accumulation of NHCP at the puff loci.[56,57] Helmsing[58,59] has demonstrated that heat-shock or ecdysone treatment results in the appearance of new protein species of 23,000 and 42,000 daltons, respectively. In contrast, Elgin and Boyd[60] noted no dramatic differences in NHCP components from heat-shocked salivary gland polytene chromatin or chromatin of normal larvae. It is quite possible that the sensitivity of the SDS electrophoresis assay systems used in the latter studies prevents the assessment of the true diversity of NHCP in puffing since only large increases in single species could be detected. However, in the work of Elgin and Boyd,[60] differences were observed in the NHCP species present in diploid chromatin vs. polytene chromatin and it was suggested by the authors that several proteins present in diploid chromatin may represent important components of constituitive heterochromatin which are underrepresented in polytene chromosomes.

In studies with another insect, Vincent and Tanguay[61] recently reported that two proteins (mol wt 34,000 and 68,000) appear in the nuclei of heat-shocked *Chironomus tetans* salivary glands. Although these components were also present in reduced amounts in the cytoplasm, the authors presented some evidence to suggest a selective nuclear localization of the proteins in response to heat treatment. Although polytene chromosomes are discussed more thoroughly in another chapter of this book, it is noteworthy that the stage-specific cycles of puffing and gene activation in polytene chromosomes provide some of the most powerful evidence for transcriptional control of gene expression during development.

The slime mold *Physarum polycephalum* represents another extremely interesting experimental model for which differentiation can be controlled and studied under laboratory conditions. LeStourgeon and Rusch[62] isolated NHCP by phenol extraction throughout the mitotic cycle of *Physarum* and at time periods during starvation-induced differentiation of the photosensitive state. While no dramatic changes in the population of NHCP were observed throughout the mitotic cycle, differentiation was accompanied with the appearance of at least four new protein bands and a loss of two bands as assessed by SDS gel electrophoresis. These findings were extended with further studies where isotope incorporation experiments indicated that the appearance of new NHCP species and quantitative changes in existing NHCP components during differentiation were the result of new protein synthesized during the periods preceding the new cell states.[63,64] However, for two proteins (46,000 and 94,000 daltons), evidence suggested that their appearance during differentiation may in part be due to the activation of mechanisms for cytoplasmic to nuclear transfer of preexisting protein. The 46,000 dalton protein has been demonstrated conclusively to be actin, which when complexed in vitro with nuclear myosin as actomyosin was bound by another protein of 34,000 daltons (having activity similar to tropomyosin), only during proliferative cell states.[65] Since the 34,000 dalton protein decreased during nonproliferative cell states, these observations pose another significant regulatory feature for nonhistones. Nuclear contractile proteins may be involved in a general inhibition or inactivation of chromatin activity via their participation in chromatin condensation. Therefore, in the above system, chromatin condensation would be inhibited during cell proliferation by the presence of the 34,000 dalton protein which in the nucleus would inhibit the activity of nuclear actomyosin.

In the higher animals, the avian erythrocyte presents an excellent system for analysis of NHCP species throughout development since relatively homogeneous populations of cells at different maturational stages and with defined transcriptional activities may be obtained. These populations include the early dividing erythroblasts (highly active in DNA, RNA, and protein synthesis), through reticulocytes (active in RNA and protein synthesis, but nondividing), to mature erythrocytes (almost inactive in RNA and protein synthesis). The progression of erythroblast to reticulocyte and erythrocyte is often termed maturation since the proerythroblast is already a cell committed to hemoglobin synthesis and the eventual acquisition of phenotypes characteristic of the mature red blood cell. In this regard the term erythrocyte differentiation is often reserved for earlier events whereby pluripotent stem cells become committed to the erythrocyte pathway and form erythroid progenitors.[66]

During maturation, the decline in transcriptional capacity occurs concomitantly with the progressive condensation of the chromatin[67-69] and with a decrease in the amount and complexity of NHCP components relative to DNA.[70-72] These events culminate to make the mature erythrocyte nucleus transcriptionally inactive.[73-75] It is noteworthy that the reduction in the mass ratio of NHCP relative to DNA from immature erythroblast to mature erythrocyte is nearly sixfold,[72-76] and occurs with little change in the amount of histone. These observations indicate a role for NHCP in erythrocyte maturation, but it remains an open question whether the loss of NHCP components is the cause or the result of nuclear inactivation.

Several laboratories have published studies in which qualitative and quantitative changes in NHCP species have been examined systematically during various stages of erythrocyte maturation. In a comparison of nonhistones extracted with buffered phenol from goose regenerating or mature erythrocyte nuclei, Shelton and Neelin[77] noted only minor variations in individual components from either development state. Sanders,[78] however, reported more dramatic differences in NHCP elements from duck erythrocyte and reticulocyte chromatin. Moreover, in this study two methods of NHCP preparation were used for either source of chromatin. Chromatin was dissociated in 3 M NaCl, DNA removed by gel filtration on BioGel® A-50, then histones separated from the residual nonhistones on SP-Sephadex®.[79] Alternatively, chromatin was dissociated in 6 M urea-0.4 M guanidine-HCl, DNA removed by ultracentrifugation, then NHCP separated from histones by ion exchange resin.[80] Not only were changes observed in the NHCP species obtained from reticulocyte and erythrocyte chromatins, but differences in the proportion and complexity of some proteins were dependent on the preparative method.[78] For example, using the NaCl and gel filtration scheme, several high molecular weight bands present in mature chromatin were absent in immature chromatin, but high molecular weight material was present in both immature and mature chromatins when the urea-ultracentrifugation-ion exchange resin preparative method was used. These data emphasize the difficulty in drawing firm conclusions as to the identity of "specific" proteins during development in that they illustrate that preparative methods influence both the presence and quantity of NHCP species obtained.

In studies of duck erythroid cell maturation, Ruiz-Carrillo et al.[72] compared changes in NHCP species throughout erythrocyte maturation. Erythroblasts were obtained from bone marrow and reticulocytes from anemic hens. Different cell types were separated according to buoyant density on albumin gradients and the NHCP complement of representative fractions then assessed with SDS gel electrophoresis. Considerable similarities in the individual components were observed throughout development, although the expected reduction in total nonhistone relative to DNA was observed from the earliest erythroblast to mature erythrocyte. Moreover, the decrease in NHCP was not evenly distributed among individual protein species during maturation since several bands were present in the same proportions throughout development and some species showed a transient increase or decrease dependent upon the maturational stage.[72] Later studies from the same laboratory investigated the synthesis of nuclear protein in duck erythroid cells using the same methods to obtain enriched populations of maturing cells.[81] Cells were pulse-labeled with ³H-leucine, separated on albumin gradients, then individual nonhistones examined on SDS gels stained for protein or counted for radioactivity incorporation. Although histone synthesis was coupled to DNA synthesis and was detectable only in erythroblasts, synthesis of nonhistones was found in all stages of maturing cells; the greatest incorporation of isotope was in erythroblasts and in general the relative amount of incorporation (corrected for the amount of protein species present), decreased progressively to the lowest levels in mature erythrocyte. Interestingly, a higher than expected ^{3}H-leucine incorporation was seen for NHCP in late polychromatic erythrocytes (Figure 1). The authors suggested that some proteins produced at this time may participate in the mechanisms responsible for nuclear inactivation of the mature erythrocyte.[81]

Vidali et al.[82] examined and compared individual components of NHCP obtained from erythroid cells of 7- or 14-day embryos or from mature hens. In 7- to 14-day embryo nuclei decreases were observed in several bands on SDS gels corresponding to molecular weights of 136 to 62,000 daltons while at the same time bands at 38,000 and 99,000 daltons exhibited net increases. The transition from fetal to adult erythrocyte was marked by reductions in the amount and intensity of a number of bands in accordance with the aforementioned studies of avian erythrocyte maturation, and also the emergence of two new bands unique to the adult cell.

Although the studies discussed here indicate that both qualitative and quantitative changes in NHCP species occur during erythrocyte maturation, it is difficult to compare different studies and assign changes to particular species since as emphasized in the studies of Sanders,[78] preparative methods markedly influence the population of NHCP components. Moreover, the presence of contaminating membrane in erythrocyte chromatin preparations can influence greatly the quantity and diversity of nonhistone species.[76,83] In studies designed to examine the effects of membrane contamination in chromatin prepared from erythrocytes at various stages of maturation, Harlow and Wells,[76] reported that simple lysis of cells in saponin followed by chromatin isolation produced chromatin heavily contaminated with putative membrane proteins. Residual membrane was excluded from chromatin if cells were treated additionally with Nonidet®-P40 and sheared with high speed rotating knives before chromatin was prepared. Nevertheless, analysis of the NHCP species in these highly purified chromatin preparations confirmed that erythrocyte maturation was accompanied by decreases in both the amount and complexity of NHCP.[76]

In general, the latter studies have examined erythrocytes at various maturational stages for changes in nonhistones solubilized from intact chromatin. However, when the bulk of nonhistones are removed from chromatin with buffered 5 *M* urea leaving only the tightly bound DNA-binding nonhistones, similar stage-specific maturational changes in these fractions can be observed.[34,84]

Recent reports have examined NHCP changes in embryonic chick skeletal muscle myoblasts during in vitro differentiation.[85,86] In these investigations, the mass ratio of NHCP relative to DNA was found to increase during the early periods of differentiation and myoblast cell fusion (24 to 66 hr), and analysis of the individual components with high resolving two-dimensional gel electrophoresis revealed that qualitative changes in some nonhistone proteins occurred during this time. In another in vitro study, explanted chick subridge mesoderm was allowed to differentiate to the cartilage phenotype for 3 day in organ culture and this process ensued with the selective loss of a high molecular weight nonhistone of about 125,000 to 150,000.[87] Systems such as those used in the latter studies[85,87] which proceed in vitro offer the opportunity to examine developmental changes in NHCP with the aid of a variety of experimental manipulations such as hormone stimulation or drug blocking experiments. Hopefully such approaches will allow the elucidation of the specific nuclear events which either cause or result from NHCP changes during development.

Recently, Teng and Teng[88] have described the biological properties and purification

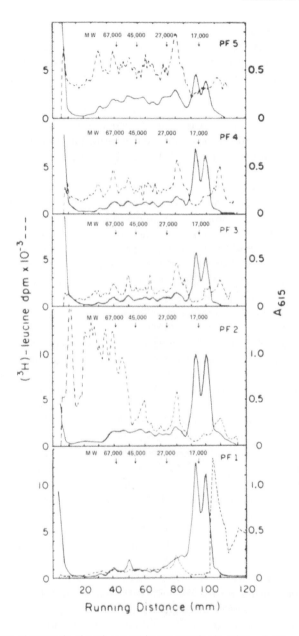

FIGURE 1.   Patterns of synthesis of nuclear nonhistone proteins in avian red cells at successive stages in maturation. A mixed population of erythroid cells was pulse-labeled with [³H]leucine for 20 min and the cells were then separated by isopycnic zonal centrifugation. Fractions correspond to the following cell types: PF-1, mature erythrocytes; PF-2, late polychromatic erythrocytes; PF-3, midpolychromatic erythrocytes; PF-4, early polychromatic erythrocytes; PF-5, early polychromatic erythrocytes plus small and large erythroblasts. Nuclei were isolated from each cell layer in the gradient. The nonhistone proteins were extracted from the dehistonized nuclei in 1% SDS at pH 8.0. About 200 µg of protein was applied to 0.6 × 10 cm cylindrical gels containing 10% polyacrylamide and 0.1% SDS at pH 7.4. The solid lines indicate the densitometric tracings of the stained protein bands. The dashed lines show the distribution of [³H]leucine activity in the nonhistone nuclear protein. Molecular weights are indicated in the scale above each panel. Note the progressive decrease in the content of nonhistone nuclear protein of molecular weights greater than 20,000 during red cell maturation. The synthesis of these proteins is also curtailed except for a selective labeling of particular peaks in late polychromatic erythrocytes (PF-2). For further details see Reference 81. (From Ruiz-Carrillo, A., Wangh, L. J., and Allfrey, V. G., *Arch. Biochem. Biophys.*, 174, 273, 1976. With permission.)

of a loosely bound tissue-specific nonhistone protein (mol wt 95,000) from hen oviduct. Interestingly, this species exhibits properties very analogous to high mobility group (HMG) proteins in that the component is easily removed from chromatin with 0.35 M NaCl and can be purified from low salt extracts using a scheme very similar to that of the HMGs.[88,89] Paradoxically however, other HMG proteins do not exhibit tissue or species specificity.[90] While the function of this protein in oviduct differentiation and maturation is unclear, it is of particular interest that the relative amount of this protein is lowest in embryonic oviduct chromatin, but with advancing age or estradiol treatment the amount increases.[91] These observations suggest a role for loosely bound NHCP in differentiation.

Assessment of mammalian tissues and cells for changes in NHCP components during differentiation and embryogenesis has also been undertaken. Changes in the composition and amount of nonhistone during perinatal development of rat liver were reported by Chytil et al.[92] Although the mass ratio of liver histones relative to DNA changed little from 19-day fetuses to adult animals, the amount of NHCP abruptly rose in 2-day animals, exhibited a slight decrease in 6-day animals, then showed another increase in 22-day animals. Only the rise in NHCP components in livers from 2-day neonates was accompanied by an increase in the template activity of chromatin as measured with *E. coli* RNA polymerase. These results have been partially confirmed in the more recent study of Guguen-Guillouzo et al.[93] These workers compared the NHCP species and parameters of nuclear protein phosphate incorporation for collagenase-dispersed hepatocytes obtained from fetal, immature, or adult rats. In accordance with Chytil et al.,[92] the amount of total NHCP rose 2 to 3 days after birth and thereafter quantitative variations in NHCP components were observed throughout maturation.

The development of two-dimensional polyacrylamide gels into a versatile and sensitive method for separation of complex mixtures of proteins is already having a tremendous impact on the analysis of nonhistones. Howe and Solter[94,95] recently used this technique to examine the synthesis of cytoplasmic and nuclear proteins in mouse embryos at different stages of preimplantation embryogenesis. They have identified several nuclear-specific, stage-dependent NHCP species present at various cleavage states. In particular, several proteins which were present at the 4 to 8 cell stage disappeared thereafter, a fate that might be expected of regulatory elements which would only be needed transiently during embryogenesis. Some of these results are shown in Figures 2 and 3.

Two-dimensional gels were also used to assess the NHCP of rat brain cortex and cerebellar neurons during normal in vivo differentiation.[96] Numerous changes were observed in the complement of nonhistones during development; especially during the developmental periods at the onset of terminal differentiation. Interestingly, two very basic, loosely bound nonhistones (35,000 and 38,000 daltons), appeared in the cortex and cerebellar nuclei coincident with the loss of proliferative capacity in these tissues.[96]

Boffa et al.[97] have analyzed changes in NHCP composition during normal differentiation of rat intestinal epithelial cells. Nuclei were isolated from normal colon using standard procedures, then separated according to buoyant density on discontinuous sucrose gradients. In this approach, nuclei which differ in buoyant density reflect differences in size, NHCP:DNA mass ratio, and DNA synthetic capacity; factors which would be expected to vary during colonic epithelial cell maturation and be related to the programming of nuclear function in various cell types. It was observed that very dense nuclei with a high DNA synthetic capacity and buoyant nuclei with a low DNA synthetic capacity differed in the amount and composition of individual NHCP components. Higher amounts of NHCP relative to DNA were found in less dense, more mature nuclei. This result correlates well with the normal histologic distribution of

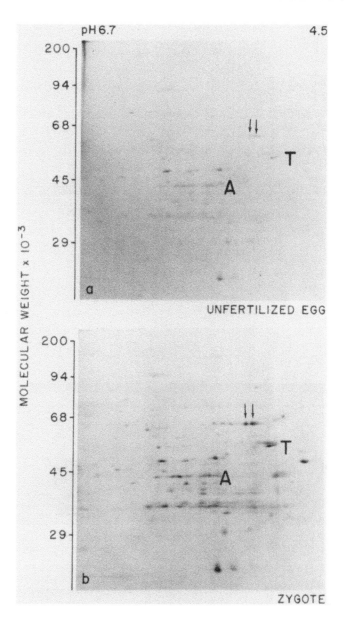

FIGURE 2. Autoradiographs of two-dimensional polyacrylamide gels of nuclear proteins from (a) unfertilized eggs, (b) zygotes, (c) 2-cell embryos, (d) 4- to 8-cell embryos, (e) blastocysts, and (f) ICM. The number of embryos used in labeling, amount of TCA-precipitable radioactivity applied to the gels, and duration of X-ray film exposure are as follows: (a) 420, 53,000 cpm, 4 months; (b) 340, 170,000 cpm, 3 months; (c) 300, 110,000 cpm, 3 months; (d) 250, 560,000 cpm, 1 month; (e) 115, 590,000 cpm, 1 month; and (f) 170, 440,000 cpm, 1 month. The arrows indicate the nuclear-specific proteins. Other details are as described in Reference 94. (From Howe, C. C. and Solter, D., *J. Embryol. Exp. Morphol.,* 52, 209, 1979. With permission.)

maturing epithelial cells in the colon, since the more mature cells, i.e., those closest to and bordering the intestinal lumen, represent terminally differentiated, mitotically arrested, but metabolically active cells.[97] These data also illustrate that differentiation

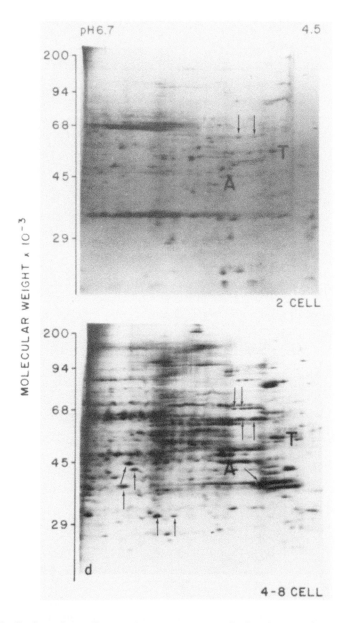

to a mitotically inactive cell state does not necessarily imply that there will be a corresponding loss of NHCP relative to DNA as was seen previously for the mature erythrocyte. In contrast to the erythrocyte, the mature epithelial cell is very active metabolically and its high NHCP:DNA mass ratio is likely a reflection of the heightened transcriptional activity necessary for production of the diversity of enzymes and products required of such a specialized cell.

Changes in NHCP species have also been examined during spermatogenesis. In the testis, spermatogenesis involves a progressive differentiation of the cells of the seminiferous epithelium from the very immature spermatogonia to the mature spermatozoon.[98] Throughout differentiation of the numerous cell types involved in spermatogenesis, synchronized changes occur in cell morphology and in the kinds and quantities of macromolecules synthesized by the cells.[98] Germinal cell nuclei contain several atyp-

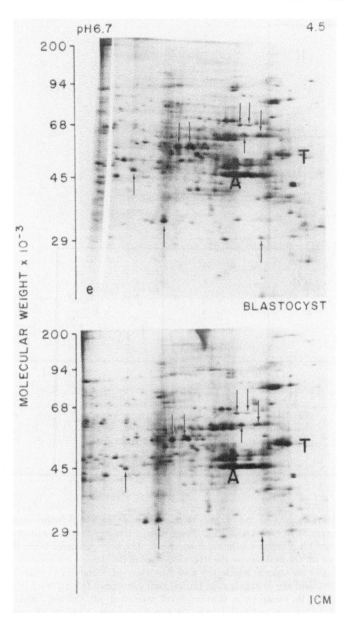

FIGURE 2.

ical chromosomal proteins[99-103] and evidence from a variety of systems indicates that some may replace somatic histones during spermatogenesis. In general these changes involve basic nuclear proteins and will not be discussed here. However, changes in the amount and synthesis of NHCP occur during development in the testes and epididymis[100,104,105] and such developmental changes are critically dependent upon pituitary gonadotrophins and testosterone.

Various alterations in the biological states of cells and tissues may be provoked by a large number of different stimuli which manifest their action through changes in gene expression. It can be argued that some of these alterations do not involve true cellular differentiation per se since in many cases a cell already committed to a particular pathway is involved. Table 1 lists representative effectors which stimulate cells to perform specific functions and in the systems listed, changes in the amount, synthesis,

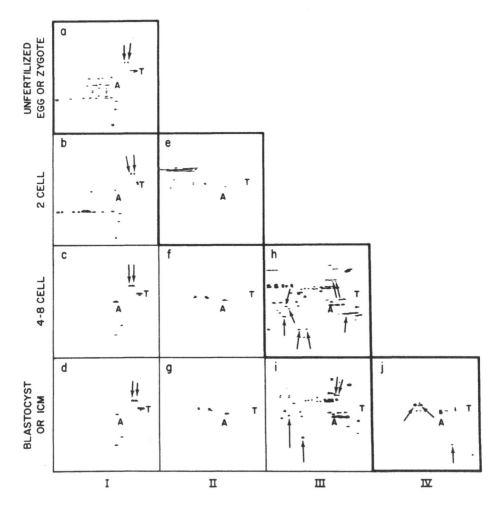

FIGURE 3.   Diagrams showing the changes in the pattern of nuclear proteins synthesized from the unfertilized egg stage up to early blastulation. Stage-specific proteins clearly identifiable in Figure 2 are shown. Nuclear-specific proteins are indicated by arrows. The heavy bordered boxes show the proteins that were observed for the first time at the stages indicated in the extreme left column. The proteins are grouped according to the stage at which they were first observed. Column 1 presents proteins first observed at the unfertilized egg or zygote stage (a) and their subsequent fate at: the 2-cell stage (b), the 4- to 8-cell stage (c), and the blastocyst or ICM (inner cell mass) stage (d). Column 2 presents proteins first observed at the 2-cell stage (e) and their fate at: the 4- to 8-cell stage (f) and the blastocyst or ICM stage (g). Column 3 presents proteins first observed at the 4- to 8-cell stage (h) and their fate at the blastocyst or ICM stage (i). Column 4 presents proteins that appeared for the first time in the blastocyst or ICM stage. In this figure, the position of actin corresponds to the top of the letter "A" and that of tubulin to the left of the letter "T". (From Howe, C. C. and Solter, D., *J. Embryol. Exp. Morphol.*, 52, 209, 1979. With permission.)

or composition of NHCP species have been observed. The listing is not meant to be inclusive and for the most part does not include known changes in nuclear enzymes such as polymerases or nucleases, nor modifications of NHCP with nuclear enzymes such as phosphokinases, methylases, or acetylases.

As a summary to the qualitative and quantitative changes in NHCP observed during differentiation and embryogenesis, several considerations should be emphasized. Stage-dependent alterations in nonhistones are seen throughout embryogenesis and differentiation, which is in accordance with the hypothesis that some of these molecules are gene regulators. In general however, reports from different laboratories using sim-

Table 1
CHANGES IN NONHISTONE CHROMOSOMAL PROTEINS IN
RESPONSE TO VARIOUS STIMULI RESULTING IN
DIFFERENTIAL GENE EXPRESSION

Cell type(s)	Stimuli	Molecular weight class affected	Ref.
Rat uterus	Estrogen	$\cong$ 20,000 daltons	106
Rat uterus	Estrogen	ND[a]	107
Rat liver	Hydrocortisone	41,000 daltons	108
Rat liver	Corticosterone	ND	109
Rat prostate	Testosterone	ND	110,111
Rat epididymis	Testosterone	Heterogeneous	104
Rat liver	Insulin	Heterogeneous	112
Rat liver	Glucagon	60,000 daltons 80,000 daltons	113
Mouse salivary gland	Isoproterenol	Heterogeneous	114
Rat ovary	Gonadotrophin	ND	115
Mouse spleen	Erythropoietin	Heterogeneous	116
Horse lymphocytes	Concanavalin A	43,000 daltons 150,000 daltons	117
Rabbit lymphocytes	Concanavalin A Anti-IG	35,000 daltons 60—80,000 daltons	118
Mouse B lympho-cytes	Bacterial Lipopolysaccharide	Heterogeneous	119,120
Rat lymphocytes	Phytohemagglutinin	Heterogeneous	121
Rat liver	Partial	45—55,000 daltons	122
Mouse liver	Hepatectomy	ND	123
W138 Human fibro-lasts	Medium change	40—60,000 daltons	124
3T6 Mouse fibro-blasts		30—80,000	124

[a] ND = Not determined.

ilar systems do not agree on individual NHCP elements involved nor on numbers and quantities of changes taking place. Possible explanations for these discrepancies might be the various preparative methods for chromatin and nonhistones in addition to differences in the resolving power of various SDS gel electrophoretic systems. Moreover, analysis of individual NHCP components using the most thorough separation procedure available, two-dimensional gel electrophoresis, does not answer questions as to how changes in single proteins might alter the extensive hierarchy of chromatin structure resulting in the pattern of expression or repression of specific genes during differentiation. A thorough discussion of this latter concept can be found in recent reviews.[8,125]

It is exceedingly difficult at the present time to estimate the amount of a regulatory nonhistone needed to alter the expression of a specific gene. If we use as models the bacterial lac repressor system,[6,126] or estimations of the number of nuclear acceptor sites for steroid hormones per cell,[127,128] about 10^4 copies per cell may be an acceptable figure. If the average protein may be considered to be about 40,000 daltons, then the average cell would contain about a femtogram of "regulatory protein". Since 1 ml of packed cells is about 5×10^8 cells, then this quantity of cells would yield about 0.7 μg of regulatory protein, which if relatively pure would be barely visible by standard methods of gel electrophoresis. These considerations emphasize the problems encountered in the characterization of probable gene regulators since sufficient amounts of material may never be available in some systems for detailed analysis of individual proteins. This discussion has been extended by some investigators to suggest that only the major nuclear enzymes and structural proteins are visualized in the analysis of NHCP preparations using conventional methods of gel electrophoresis.

B. Tissue and Cell Specificity

During cytodifferentiation and cell specialization in organogenesis large segments of the cellular DNA are transcriptionally inactivated and only those DNA sequences needed for cell-specific activities remain open for transcription. Accordingly, one would anticipate that regulatory molecules involved in the selective expression or repression of DNA would exhibit considerable heterogeneity and specificity from cell type to cell type. For these reasons, substantial effort has been expended to identify tissue-specific components among the nonhistones.

As noted in the previous section, preparative and analytical methods used for the analysis of NHCP may influence greatly the amount and composition of protein examined. Earlier reports, where investigators prepared nonhistones in bulk and compared components from different tissues using gel electrophoresis, agree that extensive homology is noted in the gel patterns of NHCP from different tissues of an organism; on the other hand, tissue-specific differences do exist.[35,129-138] These results may be interpreted to mean that common structural proteins, enzymes concerned with nucleic acid metabolism and biosynthesis, and the various enzymes of histone and nonhistone modification would be present in most tissues and thus contribute to the extensive homology observed. Moreover, since in almost all studies the heterogeneity of preparations has been examined with gel electrophoresis under denaturing conditions, differences in the primary structure between polypeptides of the same molecular weight have not been detected.

Various laboratories have fractionated chromatin in attempts to obtain preparations enriched in tissue-specific proteins. Although it is not possible here to elaborate on the various techniques utilized, most often chromatin is dissociated with salt and/or urea in combination with various buffers.[33,139,140] Using this methodology, nonhistones have been separated into protein fractions enriched in tightly associated DNA binding proteins,[141-147] DNA binding phosphoproteins,[148] loosely associated DNA binding proteins,[32,149] and various other fractions containing proteins with similar ionic and solubility properties.[30,31,136,150] Unfortunately, few studies have included a rigid examination of tissue-specific elements involved in differentiation that might be included in these fractions. However groups of nonhistones which show preferential affinity for binding to homologous DNA,[142-145] stimulation of tissue-specific in vitro DNA-dependent RNA synthesis,[20,33,146,149] or gene-specific RNA synthesis[34,151-154] have been identified. These latter functions are attributes of proteins that one would anticipate to be required of gene-specific regulatory molecules. A more thorough discussion of these concepts and the nonhistones involved can be found in other chapters of this book.

Specific antibodies raised to chromosomal components and used to examine the composition of various chromatins in different biological states have provided powerful evidence for the existence of tissue and cell-specific proteins. Antibodies offer a degree of selectivity and specificity not afforded in more conventional methods of chromosomal protein analysis such as gel electrophoresis.

Immunological approaches to chromosomal proteins began with the early studies of Messineo[155] who prepared antibodies against chromatin from normal and leukemic human white blood cells and demonstrated immunological differences between the two chromatins. Chytil and Spelsberg[156] immunized rabbits with diethylstilbestrol-stimulated chick oviduct chromatin which was dehistonized with buffered salt and urea before injection. Antisera obtained by this protocol selectively recognized components of oviduct chromatin and showed only minimal reactivity with chromatins from other chicken tissues. These earlier results have been extended and verified by a number of investigators and when chromatin or DNA-nonhistone protein complexes were used as antigens, antibodies could be elicited to chromatin nonhistone protein components specific to the species,[157,158] to the tissue,[145,156,158-160] or to normal or transformed cells[150,160-165] (for recent reviews see References 30, 31, and 163). Moreover, a major feature of antibodies against tight-binding chromosomal protein-DNA complexes is that they apparently require a specific association with homologous DNA for maximal immunoreactivity.[160-163] This latter result indicates these proteins to be attractive candidates for a function in specific gene regulation.

Of particular relevance are the studies which have used the latter techniques to detect immunologically specific changes in the nonhistones of developing tissues. In recent studies from Elgin's laboratory, antibodies were prepared against NHCP from *Drosophila* embryos or heat-shocked larvae and used to examine the in vivo distribution of NHCP by immunofluoresence in polytene chromosome squashes from *Drosophila* salivary glands.[166-168] Different NHCP components were observed to have different distributions along polytene chromosomes and some antisera prepared against proteins released from chromatin by mild DNase I digestion were seen to localize selectively on puff loci. Chytil et al.[92] observed that liver chromain from 19-day fetuses was only marginally reactive with antisera prepared against adult rat liver dehistonized chromatin. However, liver chromatin from animals of advancing age exhibited progressively greater immunoreactivity until a maximum was reached with adult liver chromatin. Some of these results are presented in Figure 4.

Similar studies have been accomplished with maturing chicken reticulocytes. Antisera prepared against reticulocyte dehistonized chromatin showed maximal immunoreactivity with chromatin from the most immature reticulocytes while chromatin from cells of increasing maturity had progressively less activity.[169] These results were interpreted to be cell-specific changes in the nuclear proteins of reticulocytes as a result of maturation, however additional evidence from the same laboratory indicates that the progressive condensation of reticulocyte chromatin (which accompanies maturation), may be in part responsible for this effect. Essentially, reticulocytes or mature erythrocytes reacted identically when immunoassays were performed in the presence of dehistonized chromatin instead of unfractionated chromatin or after prior dispersion of chromatins with sonication or dispersing agents such as dextran sulfate.[170] Nevertheless, antisera prepared against dehistonized chromatin from erythrocytes or reticulocytes exhibit remarkable tissue and species specificity and evidence suggests a high molecular weight (90,000 dalton) protein complexed with DNA to be responsible for at least part of the cell-specific immunoreactivity.[170] Further studies on the ontogeny and physicochemical properties of this component are likely to reveal important new insights into the role of NHCP in erythrocyte maturation.

IV. CONCLUSIONS

Data from numerous systems demonstrate changes in the amount and composition of nonhistones during differentiation and embryogenesis which suggests the involvement of this class of nuclear proteins in these developmental processes. While this discussion has not focused directly on the current proposed mechanisms for nonhistones in the regulation of cellular differentiation, it should be emphasized that few laboratories have isolated or identified distinct components for which definite cause and effect relationships may be established. Admittedly, the heterogeneity of the nonhistones and the technical difficulties involved in the analysis of chromatin composi-

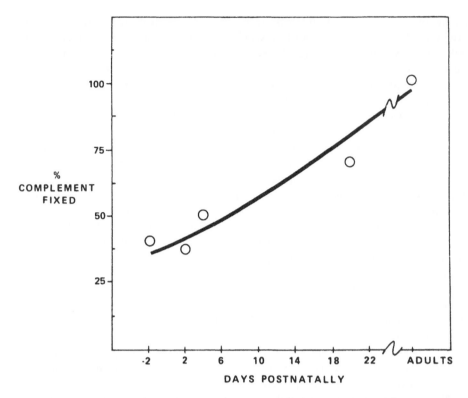

FIGURE 4.   Changes in antigenicity of acidic protein-DNA complex of the developing rat liver. From each stage of development (−2, 2, 5, 21 days old and adult) 10 μg of protein from each chromatin per assay were added to the complement fixation assays containing the antisera against adult rat liver acidic protein-DNA complex. For further details see Reference 92. (From Chytil, F., Glasser, S. R., and Spelsberg, T. C., *Dev. Biol.*, 37, 295, 1974. With permission.)

tion, structure, and function make this a very difficult task. Nevertheless, a plethora of studies indicate that nonhistones participate in almost all aspects of nuclear function. Future experimentation should establish the important temporal relationships of the various nuclear events which ultimately result in the programmed expression of specific genes during cellular differentiation.

Methods with gene cloning, restriction enzymes, and the discovery of intervening sequences in eukaryotic genes have generated an explosion of knowledge concerning the organization and expression of specific genes in the eukaryotic genome. We are now reaching a significant turning point where new methods will allow us to ask definite questions about the nonhistone regulation of gene expression during development and to unravel the mechanisms whereby protein promoters and transcriptional factors might exert influences on the genome of eukaryotic cells. In this regard the recent, elegant studies conducted with the regulation of 5S rRNA during *Xenopus* oocyte maturation have demonstrated that binding of a 40,000 dalton protein to the 5S RNA gene coding region regulates 5S rRNA production.[171,172] Gene cloning techniques together with chromatin reconstitution and identification of specific nonhistone species with highly resolving two-dimensional gel electrophoresis and antibodies provide a workable format for future efforts. It is certainly not an overstatement to suggest that undersanding the role of nonhistones in cellular differentiation will have profound implications in our understanding of biology and medicine.

ACKNOWLEDGMENT

Supported by National Cancer Institute Grant CA-26412.

REFERENCES

1. Boivin, A., Vendrely, R., and Vendrely, C., Biochimie de l'heredite. L'acide desoxyribonucleique du noyan cellulaire, dépositaire des caractères heréditaires; arguments d'orde analytique, *C. R. Acad. Sci.*, 226, 1061, 1948.
2. Mirsky, A. E. and Ris, H., Variable and constant components of chromosomes, *Nature (London)*, 163, 666, 1949.
3. Gurdon, J. B., Adult frogs derived from the nuclei of single somatic cells, *Dev. Biol.*, 4, 256, 1962.
4. Gurdon, J. B., Nuclear transplantation in amphibia and the importance of stable nuclear changes in promoting cellular differentiation, *Q. Rev. Biol.*, 38, 54, 1963.
5. McCarthy, B. J. and Hoyer, B. H., Identity of DNA and diversity of messenger RNA molecules in normal mouse, *Proc. Natl. Acad. Sci. U.S.A.*, 52, 915, 1964.
6. Jacob, F. and Monod, J., Genetic regulatory mechanisms in the synthesis of proteins, *J. Mol. Biol.*, 3, 318, 1961.
7. Hnilica, L. S., *The Structure and Biological Function of Histones*, CRC Press, Boca Raton, Fla., 1972.
8. Elgin, S. C. R. and Weintraub, H., Chromosomal proteins and chromatin structure, *Annu. Rev. Biochem.*, 44, 725, 1975.
9. Delange, R. J., Fambrough, D. M., Smith, E. L., and Bonner, J., Calf and pea histone IV III complete amino acid sequence of pea seedling histone IV: comparison with the homologous calf thymus histone, *J. Biol. Chem.*, 244, 5669, 1969.
10. Bustin, M. and Cole, R. D., Species and organ specificity in very lysine-rich histones, *J. Biol. Chem.*, 243, 4500, 1968.
11. Bustin, M. and Stollar, B. D., Immunochemical specificity in lysine-rich histone subfractions, *J. Biol. Chem.*, 247, 5716, 1972.
12. Orengo, A. and Hnilica, L. S., *In vivo* incorporation of labelled amino acids into nuclear proteins of the sea urchin embryos, *Exp. Cell Res.*, 62, 331, 1970.
13. Hohman, P., The H1 class of histone and diversity in chromosomal structure, *Subcell. Biochem.*, 5, 87, 1978.
14. Byrd, E. W., Jr. and Kasinsky, H. E., Histone synthesis during early embryogenesis in *Xenopus laevis* (South African clawed toad), *Biochemistry*, 12, 246, 1973.
15. Marushige, K. and Ozaki, H., Properties of isolated chromatin from sea urchin embryo, *Dev. Biol.*, 16, 474, 1967.
16. Adamson, E. D. and Woodland, H. R., Histone synthesis in early amphibian development: histone and DNA syntheses are not coordinated, *J. Mol. Biol.*, 88, 263, 1974.
17. Woodland, H. R., Histone synthesis during the development of *Xenopus, FEBS Lett.*, 121, 1, 1980.
18. Allfrey, V. G., Littau, V. C., and Mirsky, A. E., On the role of histones in regulating ribonucleic acid synthesis in the cell nucleus, *Proc. Natl. Acad. Sci. U.S.A.*, 49, 414, 1963.
19. Huang, R. C., Bonner, J., and Murray, K., Physical and biological properties of soluble nucleohistones, *J. Mol. Biol.*, 8, 54, 1964.
20. Paul, J. and Gilmour, R. S., Organ-specific restriction of transcription in mammalian chromatin, *J. Mol. Biol.*, 34, 305, 1968.
21. DeLange, R. J. and Smith, E. L., Histones: structure and function, *Annu. Rev. Biochem.*, 40, 279, 1971.
22. Elgin, S. C. R., Froehner, S. C., Smart, J. E., and Bonner, J., The biology and chemistry of chromosomal proteins, *Adv. Cell. Mol. Biol.*, 1, 1, 1971.
23. Bonner, J. and Garrard, W. T., Biology of the histones, *Life Sci.*, 14, 209, 1974.
24. Kornberg, R. D., Structure of chromatin, *Annu. Rev. Biochem.*, 46, 931, 1977.
25. Felsenfeld, G., Chromatin, *Nature (London)*, 271, 115, 1978.
26. Lacy, E. and Axel, R., Analysis of DNA of isolated chromatin subunits, *Proc. Natl. Acad. Sci. U.S.A.*, 72, 3978, 1975.
27. MacGillivray, A. J., The analysis of chromatin non-histone proteins, in *The Organization and Expression of the Eukaryotic Genome*, Bradbury, E. M. and Javaherian, K., Eds., Academic Press, New York, 1977, 21.
28. MacGillivray, A. J., Non-histone proteins as gene regulators? *Biochem. Soc. Trans.*, 4, 976, 1976.
29. MacGillivray, A. J. and Rickwood, D., The role of chromosomal proteins as gene regulators, in *Biochemistry of Cell Differentiation*, Vol. 9, Paul, J., Ed., University Park Press, Baltimore, 1974, 301.
30. Chiu, J. F. and Hnilica, L. S., Nuclear nonhistone proteins: chemistry and function, in *Chromatin and Chromosome Structure*, Jei Li, H. and Eckhardt, R., Eds., Academic Press, New York, 1977, 193.
31. Hnilica, L. S., Chiu, J. F., Hardy, K., and Fujitani, H., Chromosomal proteins in differentiation, in *Cell Differentiation and Neoplasia*, Saunders, G. F., Ed., Raven Press, New York, 1978, 325.

32. Kostraba, N. C. and Wang, Y. T., Inhibition of transcription *in vitro* by a non-histone protein isolated from Ehrlich ascites tumor chromatin, *J. Biol. Chem.*, 250, 8938, 1975.
33. Spelsberg, T. C., Hnilica, L. S., and Ansevin, A. T., Proteins of chromatin in template restriction III. The macromolecules in specific restriction of the chromatin DNA, *Biochim. Biophys. Acta*, 228, 550, 1971.
34. Chiu, J. F., Tsai, J. H., Sakuma, K., and Hnilica, L. S., Regulation of *in vitro* mRNA transcription by a fraction of chromosomal proteins, *J. Biol. Chem.*, 250, 9431, 1975.
35. David, A. R. and Burdman, J. A., Acidic proteins in rat brain nuclei: disc electrophoresis, *J. Neurochem.*, 15, 25, 1968.
36. Paul, J., Gilmour, R. S., More, I., MacGillivray, A. J., and Rickwood, D., Gene masking in cell differentiation: the role of nonhistone chromosomal proteins, in *Regulation of Transcription and Translation in Eukaryotes*, Bautz, E., Karlson, P., and Kersten, H., Eds., Springer-Verlag, Berlin, 1973, 31.
37. Tsai, S. Y., Harris, S. E., Tsai, M. J., and O'Malley, B. W., Effects of estrogen on gene expression in chick oviduct: the role of chromatin proteins in regulating transciption of the ovalbumin gene, *J. Biol. Chem.*, 251, 4713, 1976.
38. Tsai, S. Y., Tsai, M. J., Harris, S. E., and O'Malley, B. W., Effects of estrogen on gene expression in chick oviduct: control of ovalbumin gene expression by nonhistone proteins, *J. Biol. Chem.*, 251, 6475, 1976.
39. Jansing, R. L., Stein, J. L., and Stein, G. S., Activation of histone gene transcription by nonhistone chromosomal proteins in WI-38 human diploid fibroblasts, *Proc. Natl. Acad. Sci. U.S.A.*, 74, 173, 1977.
40. Friedman, M., Shull, K. H., and Farber, E., Highly selective *in vitro* methylation of rat liver nuclear protein, *Biochem. Biophys. Res. Commun.*, 34, 857, 1969.
41. Kish, U. and Kleinsmith, L. J., Nuclear protein kinases. Evidence for their heterogeneity, tissue specificity, substrate specificities, and differential responses to cyclic adenosine 3',5'-monophosphate, *J. Biol. Chem.*, 249, 750, 1974.
42. Kleinsmith, L. J., Acidic nuclear phosphoproteins, in *Acidic Proteins of the Nucleus*, Cameron, I. R. and Jeter, J. R., Eds., Academic Press, New York, 1974, 103.
43. Suria, S. and Liew, C. C., Isolation of nuclear acidic proteins from rat tissues. Characterization of acetylated liver nuclear acidic proteins, *Biochem. J.*, 137, 355, 1974.
44. Greenway, P. J. and Levine, D., Identification of a soluble protein methylase in chicken embryo nuclei, *Biochim. Biophys. Acta*, 350, 374, 1974.
45. Nguyen thi, M., Morris, G. E., and Cole, R. J., Gene activation during muscle differentiation and the role of nonhistone chromosomal protein phosphorylation, *Dev. Biol.*, 47, 81, 1975.
46. Hinegardner, R. T., Growth and development of the laboratory cultured sea urchin, *Biol. Bull.*, 137, 465, 1969.
47. Hill, R. J., Poccia, D. L., and Doty, P. J., Towards a total macromolecular analysis of Sea Urchin embryo chromatin, *J. Mol. Biol.*, 61, 445, 1971.
48. Gineitis A. A., Stankevičiute, J. V., and Vorab'ev, V. I., Chromatin proteins from normal vegetalized and animalized sea urchin embryos, *Dev. Biol.*, 52, 181, 1976.
49. Poccia, D. L. and Hinegardner, R. T., Developmental changes in chromatin proteins of the Sea Urchin from blastula to mature larva, *Dev. Biol.*, 45, 81, 1975.
50. Sevaljevic, L., Krtolica, K., and Konstantinovic, M., Embryonic stage-related properties of sea urchin embryo chromatin, *Biochim. Biophys. Acta*, 425, 76, 1976.
51. Cognetti, G., Settineri, D., and Spinelli, G., Developmental changes of chromatin non-histone proteins in sea urchins, *Exp. Cell Res.*, 71, 465, 1972.
52. Hnilica, L. S. and Johnson, A. W., Fractionation and analysis of nuclear proteins in Sea Urchin embryos, *Exp. Cell Res.*, 63, 261, 1970.
53. Seale, R. L. and Aronson, A. I., Chromatin-associated proteins of the developing Sea Urchin embryo, *J. Mol. Biol.*, 75, 633, 1973.
54. Elgin, S. R. and Hood, L. E., Chromosomal proteins of *Drosophila* embryos, *Biochemistry*, 12, 4984, 1973.
55. Beerman, W., *Developmental Studies on Giant Chromosomes*, Springer-Verlag, Berlin, 1972.
56. Holt, Th. K. H., Local protein accumulation during gene activation. II. Interferometric measurements of the amount of solid material in temperature induced puffs of *Drosophila hydei*, *Chromosoma (Berlin)*, 32, 428, 1971.
57. Berendes, H. D., Synthetic activity of polytene chromosomes, *Int. Rev. Cyt.*, 35, 61, 1973.
58. Helmsing, P. J. and Berendes, H. D., Induced accumulation of nonhistone proteins in polytene nuclei of *Drosophila hydei*, *J. Cell Biol.*, 50, 893, 1971.
59. Helmsing, P. J., Induced accumulation of nonhistone proteins in polytene nuclei of *Drosophila hydei*. Accumulation of proteins in polytene nuclei and chromatin of different larval tissues, *Cell Differ.*, 1, 19, 1972.

60. Elgin, S. C. R. and Boyd, J. B., The proteins of polytene chromosomes of *Drosophila hydei, Chromosoma,* 51, 135, 1975.
61. Vincent, M. and Tanguay, R. M., Heat-shock induced proteins in the cell nucleus of the *Chironomus tetans* salivary gland, *Nature (London),* 281, 501, 1979.
62. LeStourgeon, W. M. and Rusch, H. P., Nuclear acidic protein changes during differentiation in *Physarum polycephalum, Science,* 174, 1233, 1973.
63. LeStourgeon, W. M. and Rusch, H. P., Localization of nucleolar and chromatin residual acidic protein changes during differentiation in *Physarum polycephalum, Arch. Biochem. Biophys.,* 115, 144, 1973.
64. Nations, C., LeStourgeon, W. M., Magun, B. E., and Rusch, H. P., The rapid intranuclear accumulation of preexisting proteins in response to high plasmodial density in *Physarum Polycephalum, Exp. Cell Res.,* 88, 207, 1974.
65. LeStourgeon, W. M., Farer, A., Yang, Y. Z., Bertram, J. S., and Rusch, H. P., Contractile proteins: major components of nuclear and chromosome nonhistone proteins, *Biochim. Biophys. Acta,* 379, 529, 1975.
66. Laijtha, L. G. and Schofield, R., On the problem of differentiation in haemopoiesis, *Differentiation,* 2, 313, 1974.
67. Davies, H. G., Structure in nucleated erythrocytes, *J. Biophys. Biochem. Cytol.,* 9, 671, 1961.
68. Davies, H. G., Electron-microscope observations on the organization of heterochromatin in certain cells, *J. Cell Sci.,* 3, 129, 1968.
69. Kernell, A. M., Bolund, L., and Ringertz, H. R., Chromatin changes during erythropoiesis, *Exp. Cell Res.,* 65, 1, 1971.
70. Shelton, K. R., Seligy, V. L., and Neelin, J. M., Phosphate incorporation into "nuclear" residual proteins of goose erythrocytes, *Arch. Biochem. Biophys.,* 153, 375, 1972.
71. Gershey, E. L. and Kleinsmith, L., Phosphorylation of nuclear proteins in avian erythrocytes, *Biochim. Biophys. Acta,* 194, 519, 1969.
72. Ruiz-Carrillo, A., Wangh, L. J., Littau, V. C., and Allfrey, V. G., Changes in histone acetyl content and in nuclear non-histone protein composition of avian erythroid cells at different stages of maturation, *J. Biol. Chem.,* 249, 7358, 1974.
73. Williams, A. F., DNA synthesis in purified populations of avian erythroid cells, *J. Cell Sci.,* 10, 27, 1972.
74. Grasso, J. A., Woodard, J. W., and Swift, H., Cytochemical studies of nucleic acids and proteins in erythrocytic development, *Proc. Natl. Acad. Sci. U.SA.,* 50, 134, 1963.
75. Cameron, I. L. and Prescott, D. M., RNA and protein metabolism in the maturation of the nucleated chicken erythrocyte, *Exp. Cell Res.,* 30, 609, 1963.
76. Harlow, R. and Wells, J. R. E., Preparation of membrane-free chromatin bodies from avian erythroid cells and analysis of chromatin acidic proteins, *Biochemistry,* 14, 2665, 1975.
77. Shelton, K. R. and Neelin, J. M., Nuclear residual proteins from goose erythroid cells and liver, *Biochemistry,* 10, 2342, 1971.
78. Sanders, L. A., Isolation and characterization of the non-histone chromosomal proteins of developing avian erythroid cells, *Biochemistry,* 13, 527, 1974.
79. Graziano, S. L. and Huang, R. C., Chromatographic separation of chick brain chromatin proteins using a SP-Sephadex column, *Biochemistry,* 10, 4770, 1971.
80. Levy, S., Simpson, R. T., and Sober, H. A., Fractionation of chromatin components, *Biochemistry,* 11, 1547, 1972.
81. Ruiz-Carrillo, A., Wangh, L. J., and Allfrey, V. G., Selective synthesis and modification of nuclear proteins during maturation of avian erythroid cells, *Arch. Biochem. Biophys.,* 174, 273, 1976.
82. Vidali, G., Boffa, L. C., Littau, V. C., Allfrey, K. M., and Allfrey, V. G., Changes in nuclear acidic protein complement of red blood cells during embryonic development, *J. Biol. Chem.,* 248, 4065, 1973.
83. Jackson, R. C., On the identity of nuclear membrane and non-histone nuclear proteins, *Biochemistry,* 15, 5652, 1976.
84. Krajeska, W., Hrabec, E., and Klyszejko-Stefanowicz, L., Changes in DNA-binding chromosomal non-histone proteins during chicken erythroid cell maturation, *Biochimie,* 60, 211, 1978.
85. Man, T. H., Morris, G. E., and Cole, R. J., Two-dimensional gel analysis of nuclear proteins during muscle differentiation *in vitro.* I. Changes in nuclear protein content, *Exp. Cell Res.,* 126, 375, 1980.
86. Man, T. N., Morris, G. E., and Cole, R. J., Two-dimensional gel analysis of nuclear proteins during muscle differentiation *in vitro.* II. Changes in protein phosphorylation, *Exp. Cell Res.,* 126, 383, 1980.
87. Newman, S. A., Birnbaum, J. A., and Yeah, G. C. T., Loss of a non-histone chromatin protein parallels *in vitro* differentiation of cartilage, *Nature (London),* 259, 417, 1976.
88. Teng, C. T. and Teng, C. S., Immuno-biochemical studies of a non-histone chromosomal protein in embryonic and mature chick oviduct, *Biochem. J.,* 185, 169, 1980.

89. Goodwin, G. H., Sanders, C., and Johns, E. W., A new group of chromatin-associated proteins with a high content of acidic and basic amino acids, *Eur. J. Biochem.*, 38, 14, 1973.

90. Goodwin, G. H., Walker, J. M., and Johns, E. W., The high mobility group nonhistone chromosomal proteins, in *The Cell Nucleus*, Vol. 6, Busch, H., Ed., Academic Press, New York, 1978, 181.

91. Teng, C. S. and Teng, C. T., Studies on sex-organ development. Changes in chemical composition and oestradiol-binding capacity in chromatin during the differentiation of chick Mullerian ducts, *Biochem. J.*, 172, 361, 1978.

92. Chytil, F., Glasser, S. R., and Spelsberg, T. C., Alterations in liver chromatin during perinatal development of the rat, *Dev. Biol.*, 37, 295, 1974.

93. Guguen-Guillouzo, C., Tichonicky, L., and Kruh, J., Hepatocyte chromosomal non-histone proteins in developing rats, *Eur. J. Biochem.*, 95, 235, 1979.

94. Howe, C. C. and Solter, D., Cytoplasmic and nuclear protein synthesis in preimplantation mouse embryos, *J. Embryol. Exp. Morphol.*, 52, 209, 1979.

95. Howe, C., Gmur, R., and Solter, D., Cytoplasmic and nuclear protein synthesis during *in vitro* differentiation of murine ICM and embryonal carcinoma cells, *Dev. Biol.*, 74, 351, 1980.

96. Heizmann, C. W., Arnold, E. M., and Kuenzle, C. C., Fluctuations of nonhistone chromosomal proteins in differentiating brain cortex and cerebellar neurons, *J. Biol. Chem.*, 255, 11504, 1980.

97. Boffa, L. C., Vidali, G., and Allfrey, V. G., Changes in nuclear non-histone protein composition during normal differentiation and carcinogenesis of intestinal epithelial cells, *Exp. Cell Res.*, 98, 396, 1976.

98. Gondos, B., *Testicular Development in the Testes*, Vol. 4, Johnson, A. D., and Gomes, W. R., Eds., Academic Press, New York, 1977, 1.

99. Kistler, W. S., Geroch, M. E., and Williams-Ashman, H. W., Specific basic proteins from mammalian testis, *J. Biol. Chem.*, 248, 4532, 1973.

100. Branson, R. E., Grimes, S. R., Jr., Yonuschot, G., and Irvin, J. L., The histones of rat testis, *Arch. Biochem. Biophys.*, 168, 403, 1975.

101. Kistler, W. S., Noyes, C., and Heinrikson, R. L., Partial structural analysis of a highly basic low molecular weight protein from rat testis, *Biochem. Biophys. Res. Commun.*, 57, 341, 1974.

102. Platz, R. D., Grimes, S. R., Meistrich, M. L., and Hnilica, L. S., Changes in nuclear proteins of rat testis cells separated by velocity sedimentation, *J. Biol. Chem.*, 250, 5791, 1975.

103. Grimes, S. R., Jr., Meistrich, M. L., Platz, R. D., and Hnilica, L. S., Nuclear protein transitions in rat testis spermatids, *Exp. Cell Res.*, 110, 31, 1977.

104. Kadohama, N. and Turkington, R. W., Changes in acidic chromatin proteins during the hormone dependent development of rat testis and epididymis, *J. Biol. Chem.*, 249, 6225, 1974.

105. Turkington, R. W. and Majumder, G. C., Gene activation during spermatogenesis, *J. Cell Physiol.*, 85, 495, 1975.

106. Barker, K. L., Estrogen-induced synthesis of histones and a specific nonhistone protein in the uterus, *Biochemistry*, 10, 284, 1971.

107. Teng, C. S. and Hamilton, T. H., Regulation by estrogen of organ-specific synthesis of a nuclear acidic protein, *Biochem. Biophys. Res. Commun.*, 40, 1231, 1970.

108. Shelton, K. R. and Allfrey, V. G., Selective synthesis of a nuclear acidic protein in liver cells stimulated by cortisol, *Nature (London)*, 228, 132, 1970.

109. Bottoms, G. D. and Jungmann, R. A., Effects of corticosterone on phosphorylation of rat liver nuclear proteins *in vitro*, *Proc. Soc. Exp. Biol. Med.*, 144, 83, 1973.

110. Couch, R. M. and Anderson, K. M., Changes in template activity of rat ventral prostate chromatin after brief or prolonged *in vitro* exposure to androgen: evidence for a succession of chromatin "conformers", *Biochem. Biophys. Res. Commun.*, 50, 478, 1973.

111. Ahmed, K., Studies on nuclear protein phosphoproteins of rat ventral prostate: incorporation of ^{32}P from [δ – ^{32}P] ATP, *Biochim. Biophys. Acta*, 243, 38, 1971.

112. Buck, M. D. and Schauder, P., *In vivo* stimulation of ^{14}C-amino acid incorporation into nonhistone proteins in rat liver chromatin induced by insulin and cortisol, *Biochim. Biophys. Acta*, 224, 644, 1970.

113. Enea, V. and Allfrey, V. G., Selective synthesis of liver nuclear acidic proteins following glucagon administration *in vivo*, *Nature (London)*, 242, 265, 1973.

114. Stein, G. and Baserga, R., The synthesis of acidic nuclear proteins in the prereplicative phase of the isoproternol-stimulated salivary gland, *J. Biol. Chem.*, 245, 6097, 1970.

115. Jungman, R.A. and Schweppe, J. S., Mechanism of action of gonadotrophin, *J. Biol. Chem.*, 247, 5535, 1972.

116. Spivak, J. L., Effect of erythropoietin on chromosomal protein synthesis, *Blood*, 47, 581, 1976.

117. Johnson, E. M., Karn, J., and Allfrey, V. G., Early nuclear events in the induction of lymphocyte proliferation by mitogens, *J. Biol. Chem.*, 249, 4990, 1974.

118. Decker, J. M. and Marchalouis, J. J., Molecular events in lymphocyte differentiation: stimulation of nonhistone nuclear protein synthesis in rabbit peripheral blood lymphocytes by anti-immunoglobulin, *Biochem. Biophys. Res. Commun.*, 74, 584, 1977.

119. **Stott, D. L. and Williamson, A. R.,** Non-histone chromatin proteins of B-lymphocytes stimulated by lipopolysaccharide, *Biochim. Biophys. Acta,* 521, 726, 1978.

120. **Stott, D. L.,** Changes in the synthesis of non-histone chromatin proteins associated with proliferation and differentiaton of B-lymphocytes, *Biochem. Soc. Trans.,* 7, 1004, 1979.

121. **Levy, R., Levy, S., Rosenberg, S. A., and Simpson, R. T.,** Selective stimulation of nonhistone chromatin protein synthesis in lymphoid cells by phytohemagglutinin, *Biochemistry,* 12, 224, 1973.

122. **Garrard, W. T. and Bonner, J.,** Changes in chromatin proteins during liver regeneration, *J. Biol. Chem.,* 249, 5570, 1974.

123. **Kruh, J., Courtois, Y., Dastugue, B., Defer, N., Gibson, K., Kamiyama, M., and Tichonicky, L.,** Role of chromatin nonhistone proteins in the regulation of cell growth and transcription, in *Regulation of Growth and Differentiated Function in Eukaryotic Cells,* Talwar, G. P., Ed., Raven Press, New York, 1975, 139.

124. **Tsuboi, A. and Baserga, R.,** Synthesis of nuclear acidic proteins in density-inhibited fibroblasts stimulated to proliferate, *J. Cell Physiol.,* 80, 107, 1972.

125. **Alberts, B., Worcel, A., and Weintraub, H.,** On the biological implications of chromatin structure, in *The Organization and Expression of the Eukaryotic Genome,* Bradbury, E. M. and Javaherian, K., Eds., Academic Press, New York, 1977, 165.

126. **Linn, S. Y. and Riggs, A. D.,** The general affinity of *lac repressor* for *E. coli* DNA: implications for gene regulation in procaryotes and eukaryotes, *Cell,* 4, 107, 1975.

127. **Katzenellenbogen, B. S. and Gorski, J.,** Estrogen actions on synthesis of macromolecules in target cells, in *Biochemical Actions of Hormones,* Vol. 3, Litwack, G., Ed., Academic Press, New York, 1975, 187.

128. **Yamamoto, K. R. and Alberts, B. M.,** Steroid receptors: elements for modulation of eukaryotic transcription, *Annu. Rev. Biochem.,* 45, 721, 1976.

129. **Teng, C. S., Teng, C. T., and Allfrey, V. G.,** Studies of nuclear acidic proteins, *J. Biol. Chem.,* 246, 3597, 1971.

130. **Elgin, S. C. R. and Bonner, J.,** Limited heterogeneity of the major nonhistone chromosomal proteins, *Biochemistry,* 9, 4440, 1970.

131. **Wu, F. C., Elgin, S. C. R., and Hood, L. E.,** Nonhistone chromosomal proteins of rat tissues. A comparative study by gel electrophoresis, *Biochemistry,* 12, 2792, 1973.

132. **Shaw, L. M. J. and Huang, R. C. C.,** A description of two procedures which avoid the use of extreme pH conditions for the resolution of components isolated from chromatins prepared from pig cerebellar and pituitary nuclei, *Biochemistry,* 9, 4530, 1970.

133. **Loeb, J. E. and Creuzet, C.,** Electrophoretic comparsons of acidic proteins of chromatin from different animal tissues, *FEBS Lett.,* 5, 37, 1969.

134. **Barrett, T. and Gould, H. J.,** Tissue and species specificity of nonhistone chromatin proteins, *Biochim. Biophys. Acta,* 294, 165, 1973.

135. **Benjamin, W. and Gellhorn, A.,** Acidic proteins of mammalian nuclei: isolation and characterization, *Proc. Natl. Acad. Sci. U.S.A.,* 59, 262, 1968.

136. **MacGillivray, A. J., Cameron, A., Krauze, R. J., Rickwood, D., and Paul, J.,** The nonhistone proteins of chromatin. Their isolation and composition in a number of tissues, *Biochim. Biophys. Acta,* 277, 384, 1972.

137. **Yeoman, L. C., Taylor, C. W., Jordon, J. J., and Busch, H.,** Two-dimensional polyacrylamide gel electrophoresis of chromatin proteins of normal rat liver and Novikoff hepatoma ascites cells, *Biochem. Biophys. Res. Commun.,* 53, 1067, 1973.

138. **Wilhelm, J. A., Ansevin, A. T., Johnson, A. W., and Hnilica, L. S.,** Proteins of chromatin in genetic restriction. IV. Comparison of histone and nonhistone proteins of rat liver nucleolar and extranucleolar chromatin, *Biochim. Biophys. Acta,* 272, 220, 1972.

139. **Beckhor, I., Kung, G. M., and Bonner, J.,** Sequence-specific interaction of DNA and chromosomal protein, *J. Mol. Biol.,* 39, 351, 1969.

140. **Huang, R. C. C. and Huang, P. C.,** Effect of protein-bound RNA associated with chick embryo chromatin on template specificity of the chromatin, *J. Mol. Biol.,* 39, 365, 1969.

141. **Gronow, M. and Griffiths, G.,** Rapid isolation and separation of the nonhistone proteins of rat liver nuclei, *FEBS Lett.,* 15, 340, 1971.

142. **Kleinsmith, L. J., Heidema, J., and Carroll, A.,** Specific binding of rat liver nuclear proteins to DNA, *Nature (London),* 226, 1025, 1970.

143. **Van der Brock, H. W. J., Nooden, L. D., Sevall, J. S., and Bonner, J.,** Isolation, purification and fractionation of nonhistone chromosomal proteins, *Biochemistry,* 12, 229, 1973.

144. **Sevall, J. S., Cockburn, A., Savage, M., and Bonner, J.,** DNA-protein interactions of the rat liver non-histone chromosomal proteins, *Biochemistry,* 14, 782, 1975.

145. **Chiu, J. F., Wang, S., Fujitani, H., and Hnilica, L. S.,** DNA-binding chromosomal non-histone proteins. Isolation, characterization and tissue specificity, *Biochemistry,* 14, 4552, 1975.

146. **Bekhor, I. and Samal, B.,** Nonhistone chromosomal protein interaction with DNA/histone complexes, *Arch. Biochem. Biophys.,* 179, 537, 1977.

147. Pederson, T. and Bharjee, J. S., A special class of non-histone protein tightly complexed with template-inactive DNA in chromatin, *Biochemistry*, 14, 3238, 1975.
148. Kleinsmith, L. J., Specific binding of phosphorylated non-histones chromatin proteins to deoxyribonucleic acid, *J. Biol. Chem.*, 248, 5648, 1973.
149. Kostraba, N. C., Montogua, R. A., and Wang, T. Y., Study of the loosely bound nonhistone chromatin proteins, *J. Biol. Chem.*, 250, 1548, 1975.
150. Chiu, J. F., Hunt, M., and Hnilica, L. S., Tissue-specific DNA-protein complexes during azo dye hepatocarcinogenesis, *Cancer Res.*, 35, 913, 1975.
151. Barrett, T., Maryanka, D., Hamlyn, P. H., and Gould, H. J., Nonhistone proteins control gene expression in reconstituted chromatin, *Proc. Natl. Acad. Sci. U.S.A.*, 71, 5057, 1974.
152. Weisbrod, S. and Weintraub, H., Isolation of a subclass of nuclear proteins responsible for conferring a DNAase I-sensitive structure on globin chromatin, *Proc. Natl. Acad. Sci. U.S.A.*, 76, 630, 1979.
153. Gilmour, R. S. and Paul, J., Tissue-specific transcription of the globin gene in isolated chromatin, *Proc. Natl. Acad. Sci. U.S.A.*, 70, 3440, 1973.
154. Stein, G. S., Park, W., Thrall, C., Mans, R., and Stein, J., Regulation of cell cycle stage specific transcription of histone genes from chromatin by nonhistone chromosomal proteins, *Nature (London)*, 257, 764, 1975.
155. Messineo, I., Immunological differences of deoxyribonucleoproteins from white blood cells of normal and leukaemic human beings, *Nature (London)*, 190, 1122, 1961.
156. Chytil, F. and Spelsberg, T. C., Tissue differences in the antigenic properties of nonhistone protein - DNA complexes, *Nature (London) New Biol.*, 233, 215, 1971.
157. Zardi, L., Lin, J., Peterson, R. O., and Baserga, R., Specificity of antibodies to nonhistone chromosomal proteins of cultured fibroblasts, in *Control of Proliferation in Animal Cells*, Clarkson, B. and Baserga, R., Eds., Cold Spring Harbor Laboratory, New York, 1974, 729.
158. Okita, K. and Zardi, L., Immunofluorescent study of chromatin proteins in cultured fibroblasts, *Exp. Cell Res.*, 86, 59, 1974.
159. Wakabayashi, K., Wang, S., Hord, G., and Hnilica, L. S., Tissue-specific nonhistone chromatin proteins with affinity for DNA, *FEBS Lett.*, 32, 46, 1973.
160. Wakabayashi, K. and Hnilica, L. S., The immunospecificity of nonhistone protein complexes with DNA, *Nature (London) New Biol.*, 242, 153, 1973.
161. Wakabayashi, K., Wang, S., and Hnilica, L. S., Immunospecificity of nonhistone proteins in chromatin, *Biochemistry*, 13, 1027, 1974.
162. Chiu, J. F., Craddock, C., Morris, H. P., and Hnilica, L. S., Immunospecificity of chromatin nonhistone protein-DNA complexes in normal and neoplastic growth, *FEBS Lett.*, 42, 94, 1974.
163. Hnilica, L. S. and Briggs, R. C., Nonhistone protein antigens, in *Cancer Markers*, Sell, S., Ed., Humana Press, Clifton, N. J., 1980, 463.
164. Briggs, R. C., Forbes, J. T., Hnilica, L. S., Montiel, M. M., and Thor, D. E., Human granulocyte-specific nuclear antigens II. Detection of antigens in human proliferative syndromes, *J. Immunol.*, 124, 243, 1980.
165. Zardi, L., Chicken antichromatin antibodies: specificity to different chromatin fractions, *Eur. J. Biochem.*, 55, 231, 1975.
166. Silver, L. M. and Elgin, S. C. R., A method for determination of the *in situ* distribution of chromosomal proteins, *Proc. Natl. Acad. Sci. U.S.A.*, 73, 423, 1976.
167. Silver, L. M. and Elgin, S. C. R., Distribution patterns of three subfractions of *Drosophila* nonhistone chromosomal proteins: possible correlations with gene activity, *Cell*, 11, 971, 1977.
168. Mayfield, J. E., Serunian, L. A., Silver, L. M., and Elgin, S. C. R., A protein released by DNAase I digestion of *Drosophila* nuclei is preferentially associated with puffs, *Cell*, 14, 539, 1978.
169. Hardy, K., Chiu, J. E., Beyer, A. L., and Hnilica, L. S., Immunological properties of fractionated avian erythroid nuclei, *J. Biol. Chem.*, 253, 5825, 1978.
170. Krajewska, W. M., Briggs, R. C., Chiu, J. F., and Hnilica, L. S., Immunologically specific complexes of chromosomal nonhistone proteins with Deoxyribonucleic acid in chicken erythroid nuclei, *Biochemistry*, 19, 4667, 1980.
171. Engelke, D. R., Ng, S. Y., Shastry, B. S., and Roeder, R. G., Specific interaction of a purified transcription factor with an internal control region of 5S RNA genes, *Cell*, 19, 717, 1980.
172. Pelham, H. R. B. and Brown, D. D., A specific transcription factor that can bind either the 5S RNA gene or 5S RNA, *Proc. Natl. Acad. Sci. U.S.A.*, 77, 4170, 1980.

Chapter 2

DNA-BINDING NONHISTONE PROTEINS

J. Sanders Sevall

TABLE OF CONTENTS

I. INTRODUCTION

The central dogma of molecular genetics states that genetic information flows from DNA to RNA to protein. Three major processes are involved in the transmission of genetic information. The first is replication; the copying of DNA to form identical daughter molecules. The second is transcription; whereby genetic information is transcribed into messenger RNA (mRNA). The third is translation; where the genetic message is decoded by the ribosomes during protein biosynthesis (translation). It is the proteins, in one way or another, that control all of metabolism. This includes the replication and the transcription of genetic information. It is assumed that proteins can interact with the genetic information (DNA) and thus regulate the expression of such information. The entire mechanism is as yet not determined; however, this review will cover current knowledge on those nonhistone proteins that can interact with eucaryotic DNA. To limit our discussion, the histones (basic proteins of the eucaryotic nucleus) will only be briefly considered in this review.

The physical structure of eucaryotic chromosomes is firmly established with histones H3, H4, H2A, and H2B interacting in a series of dimeric complexes that lead from a H3-H4 dimer to an isologous tetramer $(H3-H4)_2$. Two H2B-H2A dimers then interact individually with the isologous tetramer $(H3-H4)_2$ to form first a hexamer and finally a heterologous octamer.[1,2] The octamer comprises the protein component of the nucleosome core particle around which is coiled, in a supercoiled configuration, ∼140 nucleotide base pairs of chromosomal DNA.[3] Neighboring nucleosomes are associated with each other by the "variable" linker DNA which affects the nucleosome to nucleosome orientation.[4-6]

Those nucleosomes that are in the region of metabolic activity can be enzymologically selected by nuclease susceptibility.[7-9] The susceptibility has been thought to be due, though not necessarily,[10,11] to the "linker region" between core nucleosomal particles.[12,13] Subsets of nucleosomes can be released via nuclease susceptibility which implies a structural internal heterogeneity possibly due to the presence of a variety of minor protein components.[13-16] Alternately, histone variants have been shown to give rise to nucleosomes which may be different in nuclease susceptibility.[17,18] Lastly, hyperacetylated histones appear in the more susceptible nucleosomes released by nuclease digestion.[19-21] Thus, several molecular mechanisms can lead to altered subfamilies of nucleosomes. Of particular interest is the relationship between the various types of nucleosome structural changes and biological activity.

Our present knowledge deals with the individual nucleosomes that comprise the template active area and those regions that may include control signals surrounding the active chromosome. However, the underlying mechanism which signals the area that will be assembled into the template active area has not yet been determined. At present, two molecular mechanisms have been postulated to "tag" specific regions of chromosomes. One is the acetylation of a specific group of a slowly metabolized family core of histones.[22] Second is a methylation of specific bases in the chromosomal DNA.[23-26] These mechanisms can be passed from generation to generation and are compatible with models of higher order chromatin structure.[27]

Changes in the expression of specific genes within nondividing cells can be assayed indirectly by immunoprecipitation[28] or directly by complementary DNA titration.[29] Induction of the functional messenger RNA by hormone action[30] or nutritional status[31] cannot be explained simply by DNA modification or histone acetylation. As yet the precise molecular mechanism is not defined but certainly would involve RNA processing, the recogniton of promoter-like sequences and/or the efficiency of promoter utilization. All these proposed mechanisms involve some type of ligand-DNA recognition mechanism. This review shall encompass current knowledge concerning the field of nonhistone protein-DNA interactions.

Those nonhistone proteins that can interact with DNA sequences can be subdivided into two broad functional classes. Those that can interact with DNA in a catalytic fashion and those that interact stoichiometrically and hence act as structural proteins.

II. CHARACTERIZED DNA-BINDING NONHISTONES

The study of DNA-binding proteins rests with proteins which act on DNA during replication,[9] recombination,[32] and gene expresson.[33] The DNA-binding proteins that have been best defined enzymatically would include DNA polymerase[34] and RNA polymerase.[35] In addition, a major class of DNA-binding proteins have been investigated that can grossly affect the physical conformation of DNA itself.[36,37] These proteins can be subclassified as helix destabilizing proteins, i.e., proteins that destabilize the double helix by binding more strongly to single-stranded DNA (ss-DNA), and the DNA-dependent ATPases that unwind DNA in a reaction driven by ATP hydrolysis.

A. Helix Destabilizing Proteins

To understand the biological properties of the metabolizing DNA-binding proteins, the properties of their bacterial counterparts will be given. The classical helix destabilizing protein is the protein synthesized from gene 32 of T₄ phage[38] and the *Escherichia coli* DNA-binding protein.[39] Both proteins can bind cooperatively with high affinity to single-stranded DNA thus destabilizing double-stranded DNA. Generally, they maintain the single-stranded configuration of DNA thus stimulating DNA synthesis by homologous DNA polymerase by lowering the rate of dissociation of the enzyme from the template.[40] Stabilizing single-stranded DNA also enhances strand reannealing between complementary DNAs which is essential in the process of replication and recombination.

This class of DNA-binding proteins can be likened to Jacob-Monod class of allosteric polypeptides. The helix destabilizing proteins lack sequence specificity but bind preferentially to single-stranded DNA (with a preference for A-T rich region in the DNA sequence). By making four basic assumptions, a theoretical analysis of this nonspecific class of DNA-binding proteins with a one-dimensional "lattice" of potential binding sites has been developed.[41] This analysis provides a simple mathematical treatment of protein binding nonspecifically with DNA which has been extended to the T₄ gene 32 protein.[42,43]

Table 1 contains a partial list of DNA-binding proteins that may function as helix destabilizing proteins in eucaryotic systems. A number of them specifically stimulate DNA polymerase — α in vitro. Other proteins stimulate their homologous DNA polymerase; while one HeLa cell DNA-binding protein stimulates β polymerase. The general consensus is the mammalian helix destabilizing proteins bind to single-stranded DNA leaving the bases exposed[44] while keeping the DNA strand in an extended conformation[45,46] and increasing the affinity of α-polymerase for the template.[9,47] Not all destabilizing proteins are involved in DNA polymerase stimulation. Two proteins isolated from meiotic cells of rat[158,162] and *Lilium*[64] definitely do not stimulate DNA polymerase. These are possibly involved in recombination and they are named R-proteins for their ability to catalyze renaturation. These proteins can *promote* helix denaturation only under conditions of helix instability. In addition, these proteins definitely are phosphorylated which affects their affinity toward single- or double-stranded DNA.

An additional class of helix destabilizing proteins is those proteins that can form nucleoprotein complexes that have a defined tertiary structure of DNA. These proteins are not the classical helix destabilizing proteins but they are included here since they seem to be involved in a packaging type of structural organization. The 9800 mol wt monomer protein of gene 5 from the filamentous bacteriophage Fd or M13 is a classic

Table 1
HELIX DESTABILIZING PROTEIN

Source	Mol wt (name)	Molecule/cell	Specificity	Helix ⇌ Coil →	Helix ⇌ Coil ←	Enzymatic effect
Calf thymus[48-50]	24,000 (UPI)	—	SS-DNA (noncooperative)	Yes	No	Stimulate α polymerase
	33,000 (Basic)	—	SS-DNA	No	No	Inhibits polymerase
	33,000-40,000 (Acidic)	—	SS-DNA	Yes	No	Stimulate α polymerase
HeLa cells[51-53] nuclei	33,000 (Basic-C1 factor)	—	SS-DNA	No	No	Stimulate α polymerase
	38,000	—	SS and DS-DNA	No	No	Stimulate β polymerase
	140,000 (Two polymerase)	—	SS DNA	—	—	Inhibits DNase
Rat cells regenerating[54-56]	24,000 (Tetramer)	10⁶	SS DNA	Yes	—	Stimulate α polymerase
Normal liver[158,162]	19,000 + 16,000 (Two polypeptides)	—	SS > DS DNA	Possibly	—	—
Normal liver	25,000 (S 25)	—	SS DNA	No	No	Inhibits DNA polymerases
Mouse cells[45]	30,000-35,000 (Phosphorylated)	—	SS-DS DNA (phosphorylated form has reduced DS affinity)	—	—	Stimulate α polymerase
Xenopus laevis[57] (oocyte)	~8,200	4 × 10¹² /Egg	SS DNA; DS DNA	—	—	Stimulate ×1 polymerase
Yeasts[59]	37,000 (Protein C)	—	SS DNA > DS DNA (noncooperative)	—	—	Yeast DNA polymerase I
Ustilago maydis[60,61]	~20,000	— (Exponential growing cells)	SS DNA	Yes	Yes	Stimulate U.m.polymerase
	46,000 and 69,000 (Glycoprotein)	10⁶	SS DNA, DS-DNA RNA (cooperative)	—	—	Solubilized DNA in 5% TCA (anti decondensation)
HSV infected cells[62]	28,000-186,000 12 DNA binding proteins	—	—	—	—	—
Rat spermatocyte[63]nuclei	33,000-35,000 (R-Protein phosphorylated)	Highest at meiosis	SS DNA	Yes -Mg⁺²	Yes + Mg⁺²	—
Lilium meiotic cells[64]	~140,000 (R-Protein phosphorylated)	Highest at meiosis	SS + DS DNA (phosphorylated)	Yes	Yes	No effect (lost when dephosphorylated)

example of such DNA-binding nonhistone.[65] The gene 5 protein is a destabilizing protein since it can bind to single strand progeny DNA and prevent synthesis of the replicative form.[66,67] It is nonspecific in that it will bind to oligonucleotides containing any of the four common bases in any sequence, although the binding may vary somewhat.[68] Its DNA-binding activity is cooperative and yields a helical rod-like structure in which there are 12 gene 5 monomers per turn of the helix.[69] Gene 5 protein oligonucleotide complex has been crystallized and may well become the first protein-DNA complex analyzed by X-ray crystallography.

Unlike helix destabilizing proteins which bind preferentially to single-stranded DNA, a low molecular weight protein from *E. coli* can bind both single-stranded and double-stranded DNA.[70] Immunologically cross-reacting proteins have been found in species of blue-green algae.[71] The protein is termed HU and it can introduce negative superhelical turns in relaxed circular SV40 DNA.[72] Its function in the bacterial genome appears similar to histones in eucaryotic genomes, i.e., it stabilizes the supercoiled configuration of double stranded DNA.[73] HMG 1 and 2 (see the chapter entitled "HMG Proteins" by Goodwin and Johns in Volume III) proteins from calf thymus nuclei, have similar properties when complexed with closed circular DNA.[74] The amino acid compositions of the *E. coli* HU and HMGs (1 and 2) are quite similar, leading to the hypothesis that the HMGs may be included in structural DNA-binding nonhistone proteins.[75] In yeast mitochondria, a histone-like packaging protein, HM, also induces supercoils when complexed with closed circular DNA and has a similar amino acid composition and DNA-binding properties as the HMGs (1 and 2).[76] The biological role of these proteins has been hypothesized to involve either transcription[77,78] or replication.[79] Certainly these proteins act as major DNA structural nonhistone proteins.

B. Topoisomerases

A second class of DNA-binding proteins listed in Table 2 modify the secondary structure and tertiary structure of DNA. Unlike the histones, the helix destabilizing proteins, and either HU or HMG, these proteins act catalytically in an energy driven reaction. These proteins may be thought of as DNA-dependent-ATPases. This group of enzymes may also be subdivided into two major subgroups: those that affect the secondary structure of DNA[80] and those that affect the tertiary structure of DNA.[81]

The proteins that affect the secondary structure of DNA are generally referred to as single-stranded DNA-dependent ATPases. One characteristic protein from bacterial sources is the *E. coli* helicase III[82] and its counterpart *rep* protein.[83] These proteins can bind to single-strand and double-strand DNA but the double-stranded binding activity is lost with the addition of ATP.[84] When bound to single-stranded DNA the enzymes can hydrolyze ATP or dATP. The DNA-binding activity is abolished by helix destabilizing proteins implying they do not bind DNA in concert.[85] Both rep and helicase III catalyze the unwinding (denaturation) of duplexed DNA but differ in respect to the mechanism of unwinding. The two proteins appear to unwind DNA by moving in opposite polarities. Each protein is attached to a single-stranded protrusion and invades the DNA helix with the help of the energy from ATP. The rep protein moves in a 3′ to 5′ direction whereas helicase III moves in a 5′ to 3′ direction. The helix destabilizing proteins can then stabilize the denatured DNA for DNA polymerase replication.

Several DNA dependent ATPases with similar activities as the bacterial enzymes have been found in eucaryotic systems (Table 2). In mouse,[86] *Lilium*,[87] human,[88,89] and calf cells,[89,101] DNA-dependent ATPases have been found. Generally, the ATPase activity increases in proliferating cells implying a role in replication.[90] The presence of the *Lilium* U protein in meiotic tissue suggests a role in recombination.

The second subgroup of DNA-binding nonhistones that possess DNA dependent ATPase activity are those enzymes that alter the tertiary structure of DNA.[36] These

Table 2
DNA DEPENDENT-ATPASES

Source	Mol wt (name)	Unwind DNA	Reduce α	Increase α coil DNA	Stimulate α DNA polymerase	Effect catenanes or knotted circles
Mouse[86]						−
Lilium[87]	130,000 (Meiotic U protein)	+	−	−	−	−
Human[88]	(ATPase)	+	−	−	+	−
Calf[89,95,101]	60,000-70,000 Topoisomerase I	−	+	−	−	−
Drosphila[96,97] embryos	Topoisomerase I	−	+	−	−	−
Yeast[98]	Topoisomerase I	−	+	−	−	−
Rat liver[100]	68,000 Topoisomerase I	−	+	−	−	−
HeLa cells[81]	Topoisomerase I	−	+	−	−	−
Mouse L cells[81]	Topoisomerase I	−	+	−	−	−
Drosphila embryos[93]	Topoisomerase II	−	+ (By 2)	+ With a H1 like protein	−	+

enzymes have the capacity to relax or coil closed circular superhelical DNA. However, not all the enzymes are ATPases; their discussion here is included for completeness. These enzymes are now classified into two major groups based on their mechanism of action. With the type I enzymes one strand of the DNA duplex is cut and rejoined at a time and appear to be uncoordinated; with the type II enzymes the breaking and rejoining of both DNA strands appear to occur in a coordinated fashion.[91-94] Three enzyme activities are associated with the Topoisomerases (I and II): (1) relaxing positive or negative topological twists in closed circular DNA; (2) unknotting topologically knotted DNA; and (3) converting DNA rings into catenated networks.

The physiological role for the topoisomerases is not completely resolved. Topoisomerase I may provide the ''swivel'' required during DNA replication to allow the DNA strands to unwind[95] and during transcription to relieve the topological constrains of RNA-DNA hybrid formation.[99] The picture is even more unclear for topoisomerase II except for several procaryotic systems[91,92] and *Drosophila* embryos.[93] ATP-dependent topoisomerase II allows one DNA helix to pass through another and may be useful in mediating the condensation and decondensation of eucaryotic chromosomes. This model would fit nicely with the radial loop theory of mitotic chromosome structure.[27] Topoisomerase II could remove the knotted conformation in metaphase chromosomes and with the help of histone H1[102-104] the topoisomerase could knot the radial loops to condensed metaphase chromosomes. Studies with *E. coli* ATP-dependent topoisomerase II (DNA gyrase) have shown that the gyrase inhibitor novobiocin specifically inhibits the transcription of ribosomal RNA in vivo while protein synthesis and the mRNA transcription are only partly affected.[105] DNA gyrase inhibitors have also been shown to inhibit the transcription of certain genes in vivo: phage N_4 DNA;[106] catabolite-sensitive operons of maltose, lactose, and galactose;[107] and the typtophan operon under control of the *lambda*-phage promoter.[108] DNA gyrase inhibitors also affect the transcription in vitro of several bacterial and plasmid genes in a cell-free system.[109] The promoters of the lactose operon, the rRNA operon rrnB and the col E_1 gene were found to be the most sensitive ones while other genes were unaffected. The effect on eucaryotic transcription is not so clearly shown. However, the transcription of an allomorphic series of templates derived from HeLa cell nuclear DNA by treatment with *E. coli* DNA gyrase or rat liver untwisting enzyme is affected.[110] Relaxed DNA is transcribed slowly by either *E. coli* RNA polymerase or wheat germ RNA polymerase II. Supertwisting (knotting) the naturally supercoiled templates with gyrase slightly inhibits transcription by the bacterial polymerase but stimulates dramatically transcription by RNA polymerase II from wheat germ. The interpretation implies that RNA polymerase II must be even more dependent on the free energy associated with negative supercoiling than is the *E. coli* polymerase for efficient transcription. It is for this reason that intact double-stranded but relaxed DNA is such a poor template for the wheat germ enzyme. In vivo, DNA is wrapped around a histone core to form nucleosomes and the isolated complex contains no free energy of supercoiling that can be released by the untwisting enzyme. In this manner, an eucaryotic gyrase may expose the template surface and so stimulate transcription.[111] Thus, it would be of interest to demonstrate the presence of a DNA-binding twisting enzyme in the eucaryotic nucleus. Alternately, the highly supercoiled DNA may act as a site for histone modification which may localize the template active region.

III. DNA-BINDING NONHISTONE PROTEINS

A. Sensitivity of DNA-Binding Activity

Nonhistone proteins by their definition are all those nucleoproteins that are not the five major histone polypeptides. These proteins are highly heterogeneous in size and isoelectric points.[112-114] The solubility properties of the nonhistone proteins in low ionic-strength buffers are such that most of these proteins can be removed from chromatin by repeated washing.[115,125] Thus, the nonhistones are now thought of as being in a dynamic flux between chromatin and the surrounding environment.[116,117] The dynamic concept of the nonhistones make analysis of the DNA-binding nonhistones extremely difficult without some way of classifying the subtypes of nonhistones. Classically, the DNA-binding nonhistones can be identified by their affinity for DNA immobilized on a stationary matrix.[118-122] Initial studies with native DNA affinity columns identified 35% of the nonhistone proteins as DNA-binding nonhistones with 3.9% having a preference toward homologous DNA.[123] However, the large amount of shared proteins between chromatin, nuclear washes, nuclear matrix, nuclear membrane, and heterogeneous RNA proteins imply a lack of compartmentalization of nuclear substructure.[124] The lack of compartmentalization dictates a more precise definition of DNA-binding nonhistones than just DNA affinity.

Additional support for this concept lies in eucaryotic cell fusion studies. Initial studies with defined aneuploid yeast strains compared in vitro the DNA-binding proteins from yeast cells with well-defined alterations in the dosage of given chromosomal DNA components.[126] These studies were done with one-dimensional gel analysis but a very small change in the amount of yeast DNA-binding proteins was ascribed to specific yeast chromosomes. To complicate matters, human-specific DNA-binding proteins in human-mouse hybrids showed no electrophoretic differences with the proteins of the individual species.[127] The major nonhistone proteins of human-hamster hybrids were indistinguishable electrophoretically from those of the parental hamster cell line with the exception of a single protein in one hybrid.[128] Although these studies are difficult to define because of the inability to define the mechanism that causes the loss of the human specific nonhistone proteins, it is clear that specific human marker proteins are not present in viable human-mouse cell hybrid. This implies that the human specific nonhistone function can be replaced by its rodent counterpart. Thus, species specific nonhistone proteins may not characterize a particular genome.

An alternative analysis of DNA-binding nonhistones would be to increase the sensitivity of detection of such proteins. This is particularly important if the implication is that only a few copies of DNA-binding nonhistones are found per genome.[129-132] The combination of several techniques increases the sensitivity of analysis of proteins that comprise a small number of molecules per nucleus.[133] The combination is based on the selection of nonhistones with DNA-binding capacity and radioactive labeling of these proteins combined with two-dimensional electrophoreses. Alternatively, a sensitive silver stain procedure of two-dimensional electrophoreses of DNA-binding nonhistones can avoid the possible chemical changes inherent in the in vitro labeling technology.[134]

This approach still is dependent upon the affinity that the DNA-binding nonhistone has toward its specific site. Analysis of the *lac* repressor affinity for nonspecific DNA-binding sites has shown the specificity of the repressor-operator binding is relatively independent of ionic strength between 0.03 to 0.2 *M*.[135,136] Specificity is being defined as the ratio of natural nonspecific DNA binding to specific DNA binding. The regulation of *lac* operon expression is the fraction of saturation of the operator by the repressor which is very much determined by the concentration of nonspecific DNA.[137,138] Hence, the DNA-affinity chromatography with the entire genome present as the immobilized substrate should isolate those proteins based on their nonspecific affinity. Alternatively, denatured DNA columns can separate those proteins that have a preference toward random coiled conformation.[139] In these studies, DNA-binding proteins in cytoplasmic and nuclear samples were isolated by using three columns in tandem orientation: cellulose, followed by double-stranded DNA cellulose, followed by single-stranded DNA cellulose. Differences in the polypeptide content of the single-stranded DNA-binding proteins and double-stranded DNA-binding proteins were observed but the localization of the proteins depends on the method of protein extraction. This supports the dynamic nature of the nonhistone-DNA interaction. In like manner the topological conformation of DNA can affect the order in which histone H1 subspecies can interact with supercoiled DNA[140] as measured by nitrocellulose filter assays. The proposed biological function of H1 on the higher order nucleofiber structure may well be dictated by the topological orientation of the DNA fiber.

In chromatographic systems,[141] the isolation of components is achieved by partitioning between two physically distinct phases that share a common interfacial boundary. Separation is achieved by moving one of these phases, the mobile or liquid phase (P_l) relative to the stationary phase (P_s). The distribution of the ligand between the two phases is a constant (partition coefficient K_{eff}) which is related to the ratio of ligand in a given volume of liquid or stationary phase:

$$K_{eff} = \frac{\text{weight of ligand in } P_s/\text{vol } P_s \text{(ligand bound)}}{\text{weight of ligand in } P_l/\text{vol } P_l \text{(ligand free)}} \tag{1}$$

The number of moles of the ligand in the stationary phase (n_s) divided by the number of moles in the liquid phase (n_l) is a constant dictated by the partition coefficient called the capacity factor (k'):

$$k' = \frac{n_s}{n_l} = \frac{\text{(ligand bound) } V_s}{\text{(ligand free) } V_l} = K_{eff} \frac{V_s}{V_l} \tag{2}$$

where V_s and V_l are the volumes of the stationary and liquid phase, respectively. The capacity factor relates the migration or retention properties of the ligand, protein in this case, to the basic properties of the system. When a protein is not retained (partitioned) with the stationary phase, its K_{eff} is zero. Hence, the capacity factor is zero.

To resolve a DNA-binding protein with apparent affinity toward the immobilized DNA on the column, the amount of liquid volume required to elute the protein is related to the partition coefficient of the protein (K_{eff}) and the capacity factor (k') of the column itself.

It is assumed that one void volume would be collected if the partition coefficient is zero. Thus the elution volume (V_e) would be equal to the volume of the liquid phase:[118]

$$V_e = V_1 \qquad\qquad (3)$$

Where the partition coefficient is greater than one the elution volume would be related to the apparent affinity of the binding protein:

$$V_e = V_1 (K_{eff} + 1) \qquad\qquad (4)$$

Since the association constant is $K_A = K_{eff}/[\text{free DNA site}]$ Equation 4 can be rearranged:

$$V_e = V_1 (1 + K_A [\text{free DNA site}]) \qquad\qquad (5)$$

Thus, the apparent elution properties of a DNA-binding protein can be monitored and quantitated.[48]

The complicating factor in this approach is nonspecific DNA affinity that all DNA-binding proteins show toward DNA. The elution volume for weak binding proteins should be small from 2 to 5 void volumes. However, the strong interactions with a DNA site in high concentration in the column may require a very large elution volume before eluting from the affinity column. An elegant technique that avoids this problem takes advantage of the inverse effect ionic strength has on both the nonspecific and specific association constants.[142] By applying a linear salt gradient to separate DNA-affinity columns that contain operator sites or nonspecific sites, ionic strength of elution of the *lac* repressor was altered by the high affinity shown toward the specific operator site. In theory, the increased ionic strength at which the repressor is eluted from the operator DNA-affinity column, when compared to the nonspecific DNA affinity column, reflects the difference between the strength of interaction of the specific site over the nonspecific site provided this difference is greater than that by which the nonspecific sites outnumber the specific sites. Hence, the success of this approach dictates a high concentration of specific DNA-binding sites on the site specific DNA affinity column.

Recombinant-DNA techniques allows the propagation of specific DNA fragments of eucaryotic genomes in bacteria vectors. This affords large amounts of material which can enrich a particular DNA-binding site. Hence, a high concentration of specific DNA-binding sites can be immobilized on a support matrix. Two DNA-binding nonhistones (DB-1 and DB-2) have been isolated from *Drosophila melanogaster* unfertilized eggs. Total DNA-binding nonhistones were isolated by urea-salt extraction of chromatin, and fractionated by Sephadex®-gel filtration and preparative gel electrophoresis. DB-1 is separated from DB-2 by preparative isoelectric focusing. DB-1 has a high affinity for a cloned fragment of nucleolar DNA.[144] By immobilizing this nucleolar clone on cellulose, DB-1 can also be purified by affinity chromatography.[143] By screening part of the *Drosophila* genome cloned in a family of plasmids, two plasmids have also been isolated that have high affinity for DB-2. Affinity chromatography with either of these plasmids has purified DB-2.[143] The purified DB-2 preferentially binds *Drosophila* DNA and comprised <0.01% of the total nonhistone chromosomal proteins.[163] The site of interaction was visualized by *in situ* hybridization to a single site on chromosome 3 in salivary gland giant chromosome. The function of both binding proteins remains to be answered. The specificity of binding suggests a regulatory role, but the use of chaotropic agents as solubilizing agents of these proteins implies a structural role in chromatin domains.

B. DNA-Binding Nonhistones Based on Solubility

Rat liver nonhistone proteins have been isolated and analyzed for DNA-binding ac-

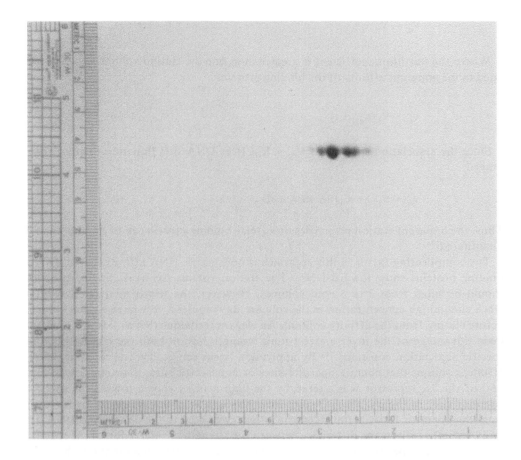

FIGURE 1.   Two-dimensional gel analysis of salt soluble nonhistone-DNA binding protein. A lyophilized DNA-binding nonhistone sample containing 150 μg of protein was electrophoresed as described by Peters and Comings.[124] Standard proteins were run alone and with the nonhistone proteins to calibrate the gel system (data not shown). The standard included bovine serum albumin, phosphorylase a, carbonic anhydrase, and soybean trypsin inhibitor. The proteins were isofocused across the top in a 3 mm × 120 mm tube gel with cathode on the left and the anode on the right. Albumin normally ran 7 to 8 cm into the isofocusing gel (pI ∼ 6.7). The second dimension was a 10% acrylamide with the nonhistone proteins running with a molecular weight of 69,000 to 70,000.

tivity. Two classes of nonhistones have been investigated, the salt soluble nonhistones[146] and the urea soluble nonhistones.[150] DNA-binding activity has been assayed by: DNA site reassociation,[145,146] mRNA driven hybridization of the DNA-binding sites,[146,147,152] enrichment of DNA restriction sites,[148,150] DNA site driven hybridization of specific gene probes,[150] and protein-binding assays of the purified binding sites.[145,149] Those nonhistones solubilized from chromatin by a 0.35-2.0 *M* sodium chloride salt extraction have been fractionated by collecting acidic proteins that are retained on the anionic exchanger DEAE cellulose. One- and two-dimensional gel analyses indicate a family of proteins are isolated that are >95% homogeneous with a 69,000 mol wt and multiple acidic isoelectric points (Figure 1). Isolation of highly sheared rat DNA fragments that are bound by the nonhistone proteins at physiologic ionic strength yields a bound DNA fraction that has been cloned in pBR322 and used as a probe for complementary sequences within the rat serum albumin gene.[164]

Labeled nonhistone-bound DNA was hybridized to nitrocellulose bound Eco RI restriction fragments of the rat serum albumin gene (Figure 2).[165] Figure 2 shows the restriction pattern and the audioradiograph with labeled total rat DNA and bound DNA as probes on the restricted fragments. The lower fragments in lanes 1 and 2

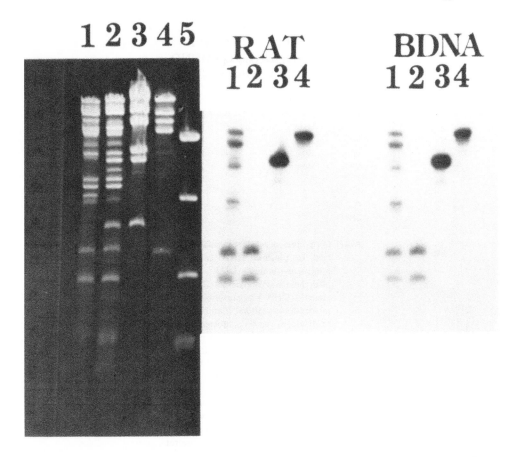

FIGURE 2. Analysis of protein-bound DNA sequences within the rat serum albumin gene. Slots 1, 2, and 3 represent a partial Eco RI restriction pattern of the 5' end (slot 1), the middle (slot 2), and the 3' end (slot 3) of the rat serum albumin.[164] The lower three restriction fragments of slots 1 and 2 represent the fourth and fifth introns of the 14 intron albumin gene and the third, fourth, and sixth exons of albumin gene. The slots labeled rat and B DNA represent Southern's analysis[165] of 1, 2, and 3 with ^{32}P labeled rat DNA and ^{32}P labeled UHP bound DNA.[146] Hence, within the third, fourth, and sixth exon of the albumin gene a repetitive DNA is bound by rat liver nonhistone proteins.

represent the D and E exon of the rat serum albumin gene.[164] Since both total rat and nonhistone-bound DNA can hybridize to these Eco RI fragments of the albumin gene, these fragments must be repetitive and are bound selectively by the nonhistone proteins. Enrichment of repetitive sequences via nonhistone-DNA interactions has been confirmed by reassociation studies where 60% of the bound DNA reassociated with a repetitive rate constant.[146] Further, the bound DNA was inserted in the Pst I site of pBR322. Nineteen colonies contained nonhistone-bound DNA inserted into pBR322 and were hybridized with total rat DNA. Table 3 confirms that 11/19 (58%) of the bound DNA clones contained repetitive sequences. Since the serum albumin messenger RNA hybridizes to single-copy DNA, the conclusion is that repetitive sequences that are bound by the nonhistone proteins are located in introns on the 5' end of the serum albumin gene. The biological function of these DNA-protein interactions has yet to be determined but current technology offers a technique for further purification of specific DNA-binding nonhistone proteins.

The urea soluble, "tight"-binding nonhistone proteins[150] from chicken or rat liver possess high affinity and preferential sequence DNA binding. In vitro DNA-binding assays have shown that the interaction is relatively ionic strength free and strongly

Table 3
HYBRIDIZATION OF [^{32}P]
TOTAL RAT DNA TO BOUND
DNA CLONES

None	Weak	Moderate	Strong
B1	B10	B9	B8
B2	B3		B11
B5	B4		B12
B7	B6		2
1	6		5
3			
4			
2B1			

Note: [^{32}P] labeled rat DNA was hybridized to plasmid DNA containing nonhistone-bound DNA inserted into the Pst I site of pBR322. The plasmid DNA was immobilized on nitrocellulose filters[165] and the hybridization was to a labeled cot = 0.15 msec. The individual plasmids were grouped by the intensity of the audioradiograph developed by overnight exposure X-ray film.

None — DNA sequence is single copy or repeated only a few times.
Weak — DNA sequence is repeated low to moderately in the rat genome.
Moderate — DNA sequence is repeated moderately in the rat genome.
Strong — DNA sequence is highly repeated in the rat genome.

prefers homologous DNA.[149] Two-dimensional gel analysis of this heterogeneous fraction of nonhistones shows similar polypeptide composition as contained in the matrix component of nucleus.[152] The homology between the tight DNA-binding nonhistones and nuclear matrix proteins suggests a functional role of these proteins in the nuclear matrix.[124,151] The function of the nuclear matrix is reviewed by Berezney in Volume IV. Currently, two different biological properties besides being an undefined nuclear skeleton have been associated with the nuclear matrix. First is the association of the SV-40,[166] or polyoma,[167] T-antigen with DNA and nuclear matrix. The T-antigen is responsible for controlling the initiation of new rounds of replication for viral[168] and possibly cellular[169] sequences. The T-antigen can introduce negative supercoils into relaxed SV-40 circles in the presence of topoisomerase I.[170] This activity appears to be sequence specific for the homologous SV-40 DNA sequence. The second biological property associated with the DNA-binding nuclear matrix is the concept that within the eucaryotic nucleus replication occurs at what is thought to be fixed sites on the nuclear matrix.[171,172] The implication is that the higher order supercoiled loops of eucaryotic chromosomes are mobile structures with DNA replication moving through the anchorage point on the nuclear matrix.

Mobile chromosomal domains may be involved in many dynamic aspects of nuclear function. For example, steroid-specific binding of estrogens and androgens to the nuclear matrix of sex steroid-responsive tissues has been demonstrated[173,175] (see Chapter 3 by Spelsberg). Although these studies do not resolve the relation of binding sites to receptor sites (the assay was indirect), these sites are saturable, high-affinity, steroid, and target tissue-specific. They are absent in the hormone withdrawn state and appear only in response to a hormonal stimuli. Another example that suggests the matrix may be involved in the biological functions of the nucleus, is the presence of tight binding nonhistones (hydrophobic in nature) that undergo a metabolic turnover with a half-life of about 5 hr.[153] These proteins stabilize the DNA double helix and appear to associate with the template-inactive fraction of DNA.[154] A tightly-bound protein isolated from *Drosophila* nuclei is preferentially associated with transcriptionally active chromosomal puffs.[155] Thus, the precise function of these proteins must await a more thorough protein analysis.

Further purification will likely result in the discovery of enzymes with overlapping activities. This is most evident for the analysis of rat liver unwinding enzyme discussed previously[54] and an alternate single-strand DNA-binding protein S-25.[158] S-25 is isolated from rat liver and prefers to bind single-stranded DNA. When bound, the protein affects the DNA helical structure inducing negative superhelical turns in the presence of the nick closing enzyme. S-25 was not observed until separated from the helix destabilizing protein of regenerating rat liver.[54] Another example is the DNA topoisomerase I and II from *Drosophila* embryos that can be separated by a salt gradient on a phosphocellulose column.[93,156] Further, the tissue from which the enzymes are isolated can help in determining the biological function of this protein. For example, the unwinding protein was isolated from regenerating rat liver but S-25 was isolated from normal liver. Hence, the unwinding protein is presumed to be involved with DNA replication. The S-25 function is not known, but its characteristics that are similar to the HMG 1 and 2 proteins imply an architectural role in chromatin structure. The *Drosophila* topoisomerases are isolated from embryos hence their role in recombination, transcription, and replication as well as chromosome segregation and condensation-decondensation is suspected.

The purified protein and its binding site will allow the analysis of the mechanism by which the enzyme functions and hence may lead to a molecular understanding of a complex function. For example, *E. coli* DNA gyrase[159] (topoisomerase II) has five different activities. It acts at specific sites[91] which can now be sequenced.[157] The mechanism of action involves a double-stranded break in the DNA[94,160] which gives rise to catenation of double-stranded DNA rings. Topoisomerase I can remove topological constrain in knotted circles or induce topological constrain with the aid of histones or HMG proteins. Via a fused-ring mechanism, it can also catalyze the formation of catenated double-stranded rings. These two activities suggest that topoisomerase I may be involved in DNA replication while topoisomerase II may be more involved with chromosome condensation-decondensation or segregation.[161]

C. Indirect Functional Assays

An indirect assay of function for DNA-binding nonhistones has been developed with in vitro initiation assays using partially purified RNA polymerases.[175] Genes which are transcribed by the class III RNA polymerases in eucaryotic cells include reiterated 5S and tRNA genes, genes encoding some small nuclear RNAs, and two adenovirus VA RNA genes expressed in productively infected human cells. The 5S RNA gene of *Xenopus laevis* has been purified[176] and has been accurately transcribed when microinjected into *Xenopus* oocytes.[177,178] This demonstrates the functionality of these purified DNAs. With the transcription of the virus-associated (VA) RNA, gene in human adenovirus 2 DNA by RNA polymerase III in a human KB cell-free extract,[179] the development of an in vitro cell-free transcription system for the *Xenopus* 5S gene was possible.[180,181] In both systems, accurate transcription is determined by sizing and sequence analysis of discrete transcripts synthesized in the in vitro system. Both homologous and nonhomologous RNA polymerase III enzymes could transcribe the DNA genes accurately. But, there is a requirement for a cellular or nuclear extract from amphibian cells for the accurate transcription of the 5S gene. Thus, an indirect assay has been developed for a factor other than purified RNA polymerase which is required for specific transcription.

The control region of the 5S gene required for accurate transcription has been mapped by enzymatically deleting a fragment of the 5S DNA from the 5' side of the gene or from the 3' side of the gene and assaying the mutant gene in the in vitro assay mix.[182,183] The gene region from nucleotides 41-87 of the 120 nucleotide 5S RNA gene contains a control region that can direct a specific transcription initiation site in the upstream DNA sequence.

The factor necessary for the accurate transcription of the cloned *Xenopus* 5S gene has been isolated from *X. laevis* ovaries.[184] This factor consists of a 37,000 dalton protein and appears to be specific for the induction of the 5S gene transcription. The site of action of the factor has been sequenced[157] within the nucleotide positions of 45 to 96 on the 5S RNA gene. The location of the protein-binding site and its control sequence correspond well with the biological function of the protein-DNA complex. Besides regulating the accurate transcription of the 5S RNA gene, the transcription factor is identical by immunological, chemical, and functional criteria to the protein associated with 5S RNA as a 7S ribonucleoprotein complex in immature *X. laevis* oocytes.[185,186] The transcription factor, which is active with somatic- and oocyte-type 5S RNA genes, appears to have a twofold function in the oocyte. First, the factor can activate the transcription of the 5S gene. Secondly, the factor can package the 5S RNA product for storage. The twofold function suggests the formal possibility that 5S RNA may feedback-inhibit (autoregulate) its own synthesis under conditions where the level of the transcription is rate-limiting, by sequestering this 5S transcription factor. This is not supported by the transcription of microinjected 5S DNA into large oocytes.[177] However, the equilibrium of the transcription factor between RNA$_{(bound)}$ and DNA$_{(free)}$ may be sufficient to dissociate the RNA protein complex to stimulate the transcription of the injected DNA. Whether or not this factor is oocyte specific is not yet determined. However, extracts of embryos or cultured *Xenopus* kidney cells can support the transcription of 5S RNA genes. However, no immunologically cross-reacting material to antitranscription factor can be detected in these extracts. This points to an alternate regulatory mechanism in somatic cells or a failure to detect low levels of transcription factor.[187] The ability of 5S RNA to inhibit 5S RNA synthesis in a germinal vesicle extract supports a direct feedback inhibition model.

Can the development of an in vitro transcriptional system be employed to assay for specific proteins that regulate the availability of a template for transcription? Several attempts to dissect the mechanisms controlling *X. laevis* 18S-28S ribosomal gene transcription with an in vitro system have proven to be aberrant. However, oocyte nuclear homogenates can direct the synthesis of the amplified ribosomal genes.[188] The interesting observation is the presence of a tightly bound component to the rDNA which cannot be removed by the addition of exogenous rDNA. Hence, polymerase I transcription in nuclear homogenates is qualitatively different from polymerase III transcription.

In vitro transcription of structural genes has been accomplished with conalbumin, ovalbumin, the early E1A, and major late adenovirus-2 genes.[189] The in vitro assay can be achieved with cloned DNA templates and purified polymerase in the presence of an S100 extract of the homologous cells. One difficulty with these studies is the lack of information on the exact site in vivo of transcription (promoter) for any eucaryotic structural gene. But, a very specific site is initiated in vitro and corresponding transcripts are found in vivo. The efficiency of the system is extremely low implying that other factors (either catalytically or structural) are probably required to account for the in vivo rates of transcription. These initial observations imply the mechanism of regulation and transcription is fundamentally different than observed in class III gene systems.

IV. SUMMARY

DNA-binding nonhistone proteins can be subdivided into two major categories: those that are catalytic in nature and those that are structural in nature. The catalytic DNA-binding nonhistones can be characterized by specific assays as DNA unwinding ATPase, topological uncoiling, DNA strand nicking, or double-strand DNA clipping activities. Undoubtedly there are additional activities that will be characterized as more defined systems are developed. Generally, the biological function of the catalytic DNA-binding nonhistone proteins is to participate in the major nuclear events of the cell such as replication, recombination, segregation, and transcription. As the catalytic mechanisms are determined, the mechanism of the individual events can be hypothesized and tested. The structural DNA-binding nonhistone proteins are still in the minority with respect to the catalytic proteins; however, their importance is currently becoming of more interest. The classic nonhistone structural proteins are the HMG proteins reviewed in Volume III by Goodwin and Johns. Basically, the HMGs are a family of nonhistones that can interact preferentially with supercoiled DNA (HMG 1 and 2) or bind to the "linker" DNA between template active nucleosomes (HMG 14 and 17). These proteins play a structural role on DNA allowing the catalytic nonhistones to carry out their metabolic function. With the hypothesis that DNA synthesis and, possibly, transcription occur at fixed sites within the nucleus, there is intense interest in those proteins that comprise the nuclear skeletal network (nuclear matrix). Hence, the finding that nuclear matrix proteins may be comprised of some sequence specific DNA-binding proteins may lead to a more definitive study in fixed DNA-binding sites on the nuclear matrix. Purification and localization of the specific binding sites on the matrix should provide information on the metabolism of the nucleus. This is an area that is just opening at the present time.

DNA-binding nonhistone proteins have been shown to interact with highly repetitive, dispersed sequences within the genome. A ubiquitous interspersed repeated sequence has been sequenced and characterized in mammalian genome.[190-192] The function of these sequences is not known. Reassociation analysis of isolated nonhistone DNA-binding sites reveal a group of highly repetitive, interspersed DNA sequences is purified from the rat genome. Highly repetitive sequences have been observed interspersed within the human α-globin,[193] β-globin genes,[194] and the rat serum albumin gene.[195] Thus a correlation between the highly repetitive DNA-binding sites and chromosomal accessibility of genetic domains may be dictated by the analysis of these specific nonhistone-DNA interactions. Localization and functional analyses are now being developed in this highly interesting area of research.

ACKNOWLEDGMENT

I acknowledge Drs. C. E. Castro and A. P. Bollon for critical reading of this manuscript and Jerry Long for clerical assistance.

REFERENCES

1. Isenberg, I., Histones, *Annu. Rev. Biochem.*, 48, 159, 1979.
2. Stein, A. and Page, D., Core histone association in high salt, *J. Biol. Chem.*, 255, 3629, 1980.
3. McGhee, J. D. and Felsenfeld, G., Nucleosome structure, *Annu. Rev. Biochem.*, 49, 1115, 1980.
4. Martin, D. F., Todd, R. D., Lang, D., Pei, P. N., and Garrard, W. T., Heterogeneity in nucleosome spacing, *J. Biol. Chem.*, 252, 8269, 1977.
5. Wilhelm, M. L. and Wilhelm, F. X., Conformation of nucleosome core particles and chromatin in high salt concentration, *Biochemistry*, 19, 4327, 1980.
6. Worcel, A. and Benyajati, C., Higher order coiling of DNA in chromatin, *Cell*, 12, 83, 1977.
7. Wu, C., Bingham, P. M., Livak, K. J., Holmgaen, R., and Elgin, S. C. R., The chromatin structure of specific genes. I. Evidence for higher order domains of defined DNA sequence, *Cell*, 16, 797, 1979.
8. Wu, C., Wong, Y. C., and Elgin, S. C. R., The chromatin structure of specific genes. II. Disruption of chromatin structure during gene activity, *Cell*, 16, 807, 1979.
9. DePhamphilis, M. L. and Wassarman, P. M., Replication of eukaryotic chromosomes: a close up of the replication fork, *Annu. Rev. Biochem.*, 49, 627, 1980.
10. Hewish, D., Internal cleavage sites of nuclease digestion, *Nuc. Acid. Res.*, 4, 1881, 1978.
11. Pospelov, V. A., Suetekova, S. B., Vorobev, V. I., Heterogeneity of chromatin subunits, *FEBS Lett.*, 74, 229, 1977.
12. Felsenfeld, G., Chromatin, *Nature (London)*, 271, 115, 1978.
13. Albright, S. C., Wiseman, J. M., Langi, R. A., and Garrard, W. T., Subunit structures of different electrophoretic forms of nucleosomes, *J. Biol. Chem.*, 255, 3673, 1980.
14. Albanase, I. and Weintraub, H., Electrophoretic separation of a class of nucleosomes enriched in HMG 14 and 17 and actively transcribed globin genes, *Nuc. Acid. Res.*, 8, 2787, 1980.
15. Mardian, J. K. W., Paton, A. E., Bunick, G. J., and Olins, D. E., Nucleosomes cores have two specific binding sites for nonhistone chromosomal proteins HMG 14 and HMG 17, *Science*, 210, 1534, 1980.
16. Nelson, P. P., Albright, S. C., Wiseman, J. M., and Garrard, W. T., Reassociation of histone H1 with nucleosomes, *J. Biol. Chem.*, 254, 11751, 1979.
17. Newrock, K. M., Alfageme, C. R., Nardi, R. V., and Cohen, L. H., Histone changes during chromatin remodeling in embryogenesis, *Cold Spring Harbor Symp. Quant. Biol.*, 42, 421, 1978.
18. Weintraub, H., Flint, S. J., Leffak, I. M., Groudine, M., and Grainges, R. M., Generation and propagation of variegated chromosome structures, *Cold Spring Harbor Symp. Quant. Biol.*, 42, 401, 1978.
19. Levy-Wilson, B., Gjerset, R. A., and McCarthy, B. J., Acetylation and phosphorylation of *Drosphila* histones, *Biochim. Biophys. Acta*, 475, 168, 1977.
20. Levy-Wilson, B. and Dixon, G. H., Partial purification of transcriptionally active nucleosomes from trout testis cells, *Nuc. Acid. Res.*, 5, 4155, 1978.
21. Davie, J. R. and Candido, E. P. M., Acetylated histone H_4 is preferentially associated with template-active chromatin, *Proc. Natl. Acad. Sci. U.S.A.*, 75, 3574, 1978.
22. Covault, J. and Chalkley, R., Identification of distinct populations of acetylated histone, *J. Biol. Chem.*, 2, 255, 9110, 1980.
23. Kuo, M. T., Mandel, J. L., and Chambon, P., DNA methylation: correlation with DNase I sensitivity of chicken ovalbumin and conalbumin chromatin, *Nuc. Acid. Res.*, 7, 2105, 1979.
24. Mandel, P. L. and Chambon, P., DNA methylation: organ specific varations in the methylation pattern within and around ovalbumin and other chicken genes, *Nuc. Acid. Res.*, 7, 2081, 1979.
25. Kaput, J. and Sneider, T. W., Methylation of somatic vs. germ cell DNAs analyzed by restriction endonuclease digestions, *Nuc. Acid. Res.*, 7, 2303, 1979.
26. Singer, J., Roberts-Ems, J., Luthardt, F. W., and Riggs, A. D., Methylation of DNA in mouse early embryos teratocarcinoma cells and adult tissues of mouse and rabbit, *Nuc. Acid. Res.*, 7, 2369, 1979.
27. Marsden, M. P. F. and Laemmli, U. K., Metaphase chromosome structure: evidence for a radial loop model, *Cell*, 17, 849, 1979.
28. Nakanishi, S., Tanabe, T., Horikawa, S., and Numa, S., Dietary and hormonal regulation of the content of acetyl coenzyme A carboxylase-synthesizing polysomes in rat liver, *Proc. Natl. Acad. Sci. U.S.A.*, 76, 2304, 1976.
29. Shafritz, D. A., Molecular hybridization probes for research in liver disease: studies with albumin cDNA, *Gastroenterology*, 77, 1335, 1979.
30. Thrall, C. L., Webster, R. A., and Spelsberg, T. C., Steroid receptor interaction with chromatin, *Cell Nucleus*, 6, 461, 1978.
31. Castro, C. E. and Sevall, J. S., Alteration of the structure and function of rat liver chromatin by nutritional factors, *Nutr. Rev.*, 38, 1, 1980.

32. **Radding, C. M.**, Genetic recombination: strand transfer and mismatch repair, *Annu. Rev. Biochem.*, 47, 847, 1978.
33. **Allfrey, V. G.**, DNA-binding proteins and transcriptional control in prokaryotic and eucaryotic systems, in *Acidic Proteins of the Nucleus*, Cameron, I. L. and Jeter, J. R., Jr., Eds., Academic Press, New York, 1974, 2.
34. **Weissbach, A.**, Eukaryotic DNA polymerases, *Annu. Rev. Biochem.*, 46, 25, 1977.
35. **Chamberlain, M. J.**, Interaction of RNA polymerase with the DNA template, in *RNA Polymerase*, Losick, R. and Chamberlain, M., Eds., Cold Spring Harbor Laboratory, Cold Spring Harbor, N.Y., 1976, 159.
36. **Champoux, J. J.**, Proteins that effect DNA conformation, *Annu. Rev. Biochem.*, 47, 449, 1978.
37. **Wang, J. C.**, Superhelical DNA, *Trends Biochem. Sci.*, August, 219, 1980.
38. **Alberts, B. M. and Frey, L.**, T$_4$ bacteriophage gene 32: a structural protein in the replication and recombination of DNA, *Nature (London)*, 227, 1313, 1970.
39. **Segal, N., Delius, H., Kornberg, T., Gefter, M. L., and Alberts, B. M.**, DNA-unwinding protein isolated from *E. coli*, *Proc. Natl. Acad. Sci. U.S.A.*, 69, 3537, 1972.
40. **Gefter, M. L.**, DNA replication, *Annu. Rev. Biochem.*, 44, 45, 1975.
41. **McGhee, J. D. and von Hippel, P. H.**, Theoretical aspects of DNA-protein interactions: co-operative and non-co-operative binding of large ligands to a one dimensional homogeneous lattice, *J. Mol. Biol.*, 86, 469, 1974.
42. **Jensen, D. E., Kelly, R. C., and von Hippel P. H.**, DNA melting proteins. II. Effects of bacteriophage T$_4$ gene 32-protein binding on the conformation and stability of nucleic acid structures, *J. Biol. Chem.*, 251, 7215, 1976.
43. **Kelly, R. and von Hippel, P. H.**, DNA melting proteins. III. Fluorescence "mapping" of the nucleic acid binding site of bacteriophage T$_4$ gene 32 protein, *J. Biol. Chem.*, 251, 7229, 1976.
44. **Otto, B. and Fanning, E.**, DNA polymerase alpha is associated with replicating SV 40 nucleoprotein complexes, *Nuc. Acid. Res.*, 5, 1715, 1978.
45. **Otto, B., Baynes, M., and Knippers, R.**, A single-strand-specific DNA-binding protein from mouse cells that stimulates DNA polymerase, *Eur. J. Biochem.*, 73, 17, 1977.
46. **Herrick, G., Delius, H., and Alberts, B.**, Single stranded structure and DNA polymerase activity in the presence of nucleic acid helix-unwinding proteins from calf thymus, *J. Biol. Chem.*, 251, 2142, 1976.
47. **Richter, A., Knippers, R., and Otto, B.**, Interaction of a mammalian single strand specific DNA binding protein with DNA polymerase alpha, *FEBS Lett.*, 91, 293, 1978.
48. **Herrick, G. and Alberts, B.**, Purification and physical characterization of nucleic acid helix-unwinding proteins from calf thymus, *J. Biol. Chem.*, 251, 2124, 1976.
49. **Herrick, G. and Alberts, B.**, Nucleic acid helix-coil transition mediated by helix-unwinding proteins from calf thymus, *J. Biol. Chem.*, 251, 2133, 1976.
50. **Kohwi-Shigematsu, T., Enomoto, T., Yamada, M. A., Nahanishi, M., and Tsuboi, M.**, Exposure of DNA bases induced by the interaction of DNA and calf-thymus DNA helix destabilizing protein, *Proc. Natl. Acad. Sci. U.S.A.*, 75, 4689, 1978.
51. **Novak, B. and Baril, E. F.**, HeLa DNA polymerase alpha activity *in vitro* specific stimulation by a non-enzymatic protein factor, *Nuc. Acid. Res.*, 5, 221, 1978.
52. **Blue, W. T. and Weissbach, A.**, Detection of a DNA polymerase beta stimulating protein in HeLa cell nuclear extracts, *Biochem. Biophys. Res. Commun.*, 84, 603, 1978.
53. **Nass, K. and Frenkel, G. D.**, A deoxyribonucleic acid binding protein from K. B. cells which inhibits deoxyribonuclease activity on single stranded DNA, *J. Biol. Chem.*, 254, 3407, 1979.
54. **Duguet, M. and de Recondo, A. M.**, A deoxyribonucleic acid unwinding protein isolated from regenerating rat liver, *J. Biol. Chem.*, 253, 1660, 1978.
55. **Patel, G. L. and Thomas, T. L.**, Some binding properties of chromatin acidic proteins with high affinity for deoxyribonucleic acid, *Proc. Natl. Acad. Sci. U.S.A.*, 70, 2524, 1973.
56. **Thomas, T. L. and Patel, G. L.**, Optimal conditions and specificity of interaction of a distinct class of nonhistone chromosomal proteins with DNA, *Biochemistry*, 15, 1481, 1976.
57. **Carrara, G., Gattoni, S., Mercanti, D., and Tocchini-Velentini, G. P.**, Purification of a DNA-binding protein from *Xenopus laevis* unfertilized eggs, *Nuc. Acid. Res.*, 4, 2855, 1977.
58. **Benbow, R. M., Breaux, C. B., Joenje, H., Krass, M. R., Lennox, R. W., Nelson, E. M., Wang, N. S., and White S. H.**, Eukaryotic DNA replication: the first steps toward a multienzyme system from *Xenopus laevis*, *Cold Spring Harbor Symp. Quant. Biol.*, 43, 597, 1979.
59. **Chang, L. M. S., Lurie, K., and Plevani, P.**, A stimulatory factor for yeast DNA polymerase, *Cold Spring Harbor Symp. Quant. Biol.*, 43, 587, 1979.
60. **Yarranton, G. T., Moore, P. D., and Spanos, A.**, The influence of DNA binding protein on the substrate affinities of DNA polymerase from *Ustilago maydis*: one polymerase implicated in both DNA replication and repair, *Mol. Gen. Genet.*, 145, 215, 1976.

61. Unran, P., Champ, D. R., Young, J. L., and Grant, C. E., Nucleic acid-binding glycoproteins which solubilize nucleic acids in dilute acid, *J. Biol. Chem.*, 255, 614, 1980.

62. Purifoy, D. J. M. and Powell, K. L., DNA-binding proteins induced by herpes simplex virus type 2 in HEp-2 cells, *J. Virol.*, 19, 717, 1976.

63. Mather, J. and Hotta, Y., A phosphorylatable DNA-binding protein associated with a liproprotein fraction from rat spermatocyte nuclei, *Exp. Cell Res.*, 109, 181, 1977.

64. Hotta, Y. and Stern, H., The effect of dephosphorylation on the properties of a helix-destabilizing protein from meiotic cells and its partial reversal by a protein kinase, *Eur. J. Biochem.*, 95, 31, 1979.

65. McPherson, A., Wang, A. H. J., Jurnak, F. A., Molineux, I., Kolpak, F., and Rich, A., X-ray diffraction studies on crystalline complexes of gene 5 DNA-unwinding protein with deoxyoligonucleotides, *J. Biol. Chem.*, 255, 3174, 1980.

66. Mazur, B. J. and Zinder, N. D., The role of gene V protein in f1 single-strand synthesis, *Virology*, 68, 490, 1975.

67. Nakashima, Y., Dunker, A. K., Marvin, D. A., and Konigsberg, W., The amino acid sequence of a DNA binding protein, the gene 5 product of filamentous bacteriophage, *FEBS Lett.*, 40, 290, 1974.

68. Anderson, R. A., Nakashima, Y., and Coleman, J. E., Chemical modifications of functional residues of fd gene 5 DNA-binding protein, *Biochemistry*, 14, 907, 1975.

69. Alberts, B., Frey, L., and Delius, H., Isolation and characterization of gene 5 product of filamentous bacterial viruses, *J. Mol. Biol.*, 68, 139, 1972.

70. Rouviere-Yaniv, J., Gross, F., Haselkorn, R., and Reiss, C., Histone-like proteins in prokaryotic organisms and their interaction with DNA, in *The Organization and Expression of the Eucaryotic Genome*, Bradbury, E. M., and Javaherian, K., Eds., Academic Press, New York, 1977, 211.

71. Haselkorn, R. and Rouviere-Yaniv, J., Cyanobacterial DNA-binding protein related to *Escherichia coli* HU, *Proc. Natl. Acad. Sci. U.S.A.*, 73, 1917, 1976.

72. Rouviere-Yaniv, J., Yaniv, M., and Germond, J. E., *E. coli* DNA binding protein HU forms nucleosome-like structure with circular double-stranded DNA, *Cell*, 17, 265, 1979.

73. Wang, J. C., Superhelical DNA, *Trends Biochem. Sci.*, August, 219, 1980.

74. Mathis, D. J., Kindelis, A., and Spadafora, C., HMG proteins (1 and 2) form beaded structures when complexed with closed circular DNA, *Nuc. Acid. Res.*, 8, 2577, 1980.

75. Spiker, S., Mardian, J. K. W., and Isenberg, I., Chromosomal HMG proteins occur in three eucaryotic kingdoms, *Biochem. Biophys. Res. Commun.*, 82, 129, 1978.

76. Caron, F., Jacq, C., and Rouviere-Yaniv, J., Characterization of a histone-like protein extracted from yeast mitochondria, *Proc. Natl. Acad. Sci. U.S.A.*, 76, 4265, 1979.

77. Vidali, G., Boffa, L. C., and Allfrey, V. G., Selective release of chromosomal proteins during limited DNAase 1 digestion of avion erythrocyte chromatin, *Cell*, 12, 409, 1977.

78. Levy, W. B., Wong, N. C. W., and Dixon, G. H., Selective association of the trout-specific H6 protein with chromatin regions susceptible to DNase I and DNase II, *Proc. Natl. Acad. Sci. U.S.A.*, 74, 2810, 1977.

79. Bidney, D. L. and Reeck, G. R., Purification from cultured hepatoma cells of two nonhistone chromatin proteins with preferential affinity for single stranded DNA, *Biochem. Biophys. Res. Commun.*, 85, 1211, 1978.

80. Alberts, B. and Sternglanz, R., Recent excitement in the DNA replication problem, *Nature (London)*, 269, 655, 1977.

81. Vosberg, H. P., Grossman, L. I., and Vinograd, J., Isolation and partial characterization of the relaxation protein from nuclei of cultured mouse and human cells, *Eur. J. Biochem.*, 55, 79, 1975.

82. Yarranton, G. J., Das, R. H., and Gefter, M. L., Enzyme catalyzed DNA unwinding, *J. Biol. Chem.*, 254, 11997, 1979.

83. Scott, J. F. and Kornberg, A., Purification of the *rep* protein of *Escherichia coli*, *J. Biol. Chem.*, 253, 3292, 1978.

84. Das, R. H., Yarranton, G. T., and Gefter, M. L., Enzyme-catalyzed DNA unwinding, *J. Biol. Chem.*, 255, 8069, 1980.

85. Yarranton, G. T., Das, R. H., and Gefter, M. L., Enzyme-catalyzed DNA unwinding, *J. Biol. Chem.*, 254, 12002, 1979.

86. Hachmann, H. J. and Lezuis, A. G., An APTase depending of the presence of single-stranded DNA from mouse myeloma, *Eur. J. Biochem.*, 61, 325, 1976.

87. Hotta, Y. and Stern, H., DNA unwinding protein from meiotic cells of lilium, *Biochemistry*, 17, 1872, 1972.

88. Cobianchi, F., Riva, S., Mastromei, G., Spardari, S., Pedrali-Noy, G., and Falaschi, A., Enhancement of the rate of DNA polymerase-alpha activity on duplex DNA by a DNA-binding protein and DNA-dependent ATPase of mammalian cells, *Cold Spring Harbor Symp. Quant. Biol.*, 43, 639, 1979.

89. Otto, B., DNA-dependent ATPases in concanavalin A stimulated lymphocytes, *FEBS Lett.*, 79, 175, 1977.

90. Mattern, M. R. and Painter, R. B., Dependence of mammalian DNA replication on DNA supercoiling. II. Effects of novobiocin on DNA synthesis in CHO cells, *Biochim. Biophys. Acta*, 563, 306, 1979.

91. Morrison, A., Higgins, N. P., and Cozzarelli, N. R., Interaction between DNA gyrase and its cleavage site on DNA, *J. Biol. Chem.*, 255, 2011, 1980.

92. Liu, L. F., Liu, C. C., and Alberts, B. N., Type II DNA topoisomerases: enzymes that can unknot a topologically knotted DNA molecule via a reversible double-strand break, *Cell*, 19, 697, 1980.

93. Hsieh, T. and Bruttag, D., ATP-dependent DNA topoisomerase from *D. melanogaster*, reversibly catenates duplex DNA rings, *Cell*, 21, 115, 1980.

94. Mizuuchi, K., Fisher, L. M., O'Dea M. H., and Gellert, M., DNA-gyrase action involves the introduction of transient double-strand breaks into DNA, *Proc. Natl. Acad. Sci. U.S.A.*, 77, 1847, 1980.

95. Champoux, J. J. and Dulbecco, R., An activity from mammalian cells that untwists superhelical DNA, a possible swivel, *Proc. Natl. Acad. Sci. U.S.A.*, 69, 143, 1972.

96. Baase, W. A. and Wang, J. C., A DNA relaxing enzyme isolated from *Drosophilia* embryos, *Biochemistry*, 13, 4299, 1974.

97. Keller, W. and Wendel, I., Steptwise relaxation of supercoiled SV 40 DNA, *Cold Spring Harbor Symp. Quant. Biol.*, 39, 199, 1975.

98. Durnford, J. M. and Champoux, J. J., The untwisting enzyme from *Saccharomyces cerevisiae*. Partial purification and characterization, *J. Biol. Chem.*, 253, 1086, 1978.

99. Warner, C. K. and Schaller, H., RNA-polymerase-promotor complex stability on supercoiled and relaxed DNA, *FEBS Lett.*, 74, 215, 1977.

100. Champoux, J. J. and McConaughy, B. L., Purification and characterization of the DNA untwisting enzyme from rat liver, *Biochemistry*, 15, 4638, 1976.

101. Pullybank, D. E. and Morgan, A. R., Partial Purification of "w" protein from calf thymus, *Biochemistry*, 14, 5205, 1975.

102. Hsiang, M. W. and Cole, R. D., Structure of histone H1-DNA complex; effect of histone-H1 on DNA condensation, *Proc. Natl. Acad. Sci. U.S.A.*, 74, 4852, 1977.

103. Renz, M., Nehls, P., and Hozier, J., Histone H1 involvement in the structure of the chromatin fiber, *Cold Spring Harbor Symp. Quant. Biol.*, 42, 245, 1977.

104. Muller, U., Zentgraf, H., Eichen, I., and Keller, W., Higher order structure of SV 40 chromatin, *Science*, 201, 406, 1978.

105. Oostra, B. A., Geert, A. B., and Gruber, M., Involvement of DNA gyrase in the transcription of ribosomal DNA, *Nuc. Acid. Res.*, 8, 4235, 1980.

106. Falco, S. C., Zivin, R., and Rothmann-Denes, Novel template requirements of N4 virion RNA polymerase, *Proc. Natl. Acad. Sci. U.S.A.*, 75, 3220, 1978.

107. Sanzey, B., Modulation of gene expression by drugs affecting DNA gyrase, *J. Bacteriol.*, 138, 40, 1979.

108. Smith, C. L., Kubo, M., and Imamoto, F., Promotor-specific inhibition of transcription by antibiotics which act on DNA gyrase, *Nature (London)*, 275, 420, 1978.

109. Yang, H. L., Heller, K., Gellert, M., and Zubay, G., Differential sensitivity of gene expression *in vitro* to inhibitors of DNA gyrase, *Proc. Natl. Acad. Sci. U.S.A.*, 76, 3304, 1979.

110. Akrigg, A. and Cook, P. R., DNA gyrase stimulates transcription, *Nuc. Acid. Res.*, 8, 845, 1980.

111. Cook, P. R., Hypothesis on differentiation and the inheritance of gene superstructure, *Nature (London)*, 245, 23, 1973.

112. Garrard, W. T., Pearson, W. R., Wake, S. K., and Bonner, J., Stoichiometry of chromatin proteins, *Biochem. Biophys. Res. Commun.*, 58, 50, 1974.

113. Peterson, J. L. and McConkey, E. H., Nonhistone chromosomal proteins from HeLa cells. A survey by high resolution, two-dimensional electrophoresis, *J. Biol. Chem.*, 251, 548, 1976.

114. Murphy, R. F. and Bonner, J., Alakline extraction of non-histone proteins from rat liver chromatin, *Biochim. Biophys. Acta*, 405, 62, 1975.

115. Comings, D. E. and Tack, L. O., Nonhistone proteins. The effect of nuclear washes and comparison of metaphase and interphase chromatin, *Exp. Cell Res.*, 82, 175, 1973.

116. Bhorjee, J. S. and Pederson, T., Chromatin: its isolation from cultured mammalian cells with particular reference to contamination by nuclear ribonucleoprotein particles, *Biochemistry*, 12, 2766, 1973.

117. Pederson, T., Gene activation in eukaryotes: are nuclear acidic proteins the cause or effect?, *Proc. Natl. Acad. Sci. U.S.A.*, 71, 617, 1974.

118. Alberts, B. and Herrick, G., DNA-cellulose chromatography in *Methods in Enzymology*, Vol. 21, Part D, Grossman, L. and Moldave, K., Eds., Academic Press, New York, 1971, 198.

119. Poonian, M. S., Schlabach, A. J., and Weissbach, A., Covalent attachment of nucleic acids to agarose for affinity chromatography, *Biochemistry*, 10, 424, 1971.

120. Schaller, H., Nüsslein, C., Bonhoeffer, F. J., Kurz, C., and Nietzschmann, I., Affinity chromatography of DNA-binding enzymes on single-stranded DNA agarous columns, *Eur. J. Biochem.*, 26, 474, 1972.

121. Rickwood, D., An improved method for the insolubilization of DNA for affinity chromatography, *Biochim. Biophys. Acta,* 269, 47, 1972.

122. Noyes, B. E. and Stark, G. R., Nucleic acid hybridization using DNA covalently coupled to cellulose, *Cell,* 5, 301, 1975.

123. van den Brock, H. W. J., Noodin, L. D., Sevall, J. S., and Bonner, J., Isolation, purification and fractionation of nonhistone chromosomal proteins, *Biochemistry,* 12, 229, 1973.

124. Peters, K. E. and Comings, D. E., Two-dimensional gel electrophoresis of rat liver nuclear washes, nuclear matrix and hnRNA proteins, *J. Cell Biol.,* 86, 135, 1980.

125. Bidney, D. L. and Reeck, G. R., Analysis of the effectiveness of sodium chloride in dissociating non-histone chromatin proteins of culture hepatoma cells, *Biochim. Biophys. Acta,* 521, 753, 1978.

126. Finkelstein, D. B. and Butow, R. A., DNA-binding proteins related to the dosage of specific yeast chromosomes, *Biochem. Biophys. Res. Commun.,* 66, 1365, 1975.

127. Jost, E., Lennox, R., and Harris, H., Affinity chromatography of DNA-binding proteins from human, marine and man-mouse hybrid cell lines, *J. Cell Sci.,* 18, 41, 1975.

128. Lydersen, B. K., Kao, F. T., and Pettijohn, D., Expression of gene coding for non-histone chromosomal proteins in human-Chinese hamster cell hybrids, *J. Biol. Chem.,* 255, 3002, 1980.

129. MacGillivray, A. J. and Rickwood, D., The heterogeneity of mouse-chromatin nonhistone proteins as evidenced by two-dimensional polyacrylamide gels, *Eur. J. Biochem.,* 41, 181, 1974.

130. Man, N. T., Morris, G. E., and Cole, R. J., Two-dimensional gel analysis of nuclear proteins during muscle differentiation, *Exp. Cell. Res.,* 121, 375, 1980.

131. Teng, C. S., Teng, C. T., and Allfrey, V. G., Studies on nuclear acidic proteins, *J. Biol. Chem.,* 246, 3597, 1971.

132. Kostraba, N. C. and Wang, T. Y., Transcriptional transformation of walker tumor chromatin by nonhistone proteins, *Cancer Res.,* 32, 2348, 1972.

133. Winter, H., Alonso, A., and Goerttler, K., A sensitive assay system for detection of rare chromosomal proteins with DNA-binding proteins, *Anal. Biochem.,* 105, 39, 1980.

134. Switzer, R. C., III, Merril, C. R., and Shifrin, S., A highly sensitive silver stain for detecting proteins and peptides in polyacrylamide gels, *Anal. Biochem.,* 98, 231, 1979.

135. Lin, S. and Riggs, A. D., *Lac* repressor affinity for nonoperator DNA, *J. Mol. Biol.,* 72, 671, 1972.

136. Lin, S. and Riggs, A. D., General affinity of *lac* repressor for *E. coli* DNA: implications for gene regulation in procaryotes and eukaryotes, *Cell,* 4, 107, 1975.

137. von Hippel, P. H., Revzin, A., Gross, C. A., and Wang, A. C., Nonspecific DNA binding of genome regulating proteins as a biological control mechanism, I. The *lac* operon; equilibrium aspects, *Proc. Natl. Acad. Sci. U.S.A.,* 71, 4808, 1974.

138. Yamomoto, K. R. and Alberts, B., The interaction of estradiol-receptor protein with the genome: an argument for the existence of undetected specific sites, *Cell,* 4, 301, 1975.

139. Miller, M. R. and Shinaberry, R. G., Subcellular location of DNA-binding proteins in baby hamster kidney cells, *J. Biol. Chem.,* 255, 8443, 1980.

140. Welch, S. L. and Cole, R. D., Differences among subfractions of H1 histone in retention of linear and superhelical DNA on filters, *J. Biol. Chem.,* 255, 4516, 1980.

141. Regnier, F. E. and Gooding, K. M., High-performance liquid chromatography of proteins, *Ann. Biochem.,* 103, 1, 1980.

142. Herrick, G., Site-specific DNA affinity chromatography of the *lac* repressor, *Nuc. Acid. Res.,* 8, 3721, 1980.

143. Weideli, H. and Gehring, W. J., A new method for the purification of DNA-binding proteins with sequence specificity, *Eur. J. Biochem.,* 104, 5, 1980.

144. Weideli, H., Scheldi, P., Artavanis-Tsakonas, S., Steward, R., Yuan, R., and Gehring, W. J., Purification of a protein from unfertilized eggs of *Drosophilia* with specific affinity for a defined DNA sequence and cloning of this sequence in plasmids, *Cold Spring Harbor Symp. Quant. Biol.,* 42, 693, 1978.

145. Jagodzinski, L. L., Chilton, J. C., and Sevall, J. S., DNA-binding nonhistone proteins: DNA site reassociation, *Nuc. Acid. Res.,* 5, 1487, 1978.

146. Jagodzinski, L. L., Castro, C. E., Sherrod, P., Lee, D., and Sevall, J. S., Reassociation kinetics of nonhistone-bound DNA sites, *J. Biol. Chem.,* 254, 3038, 1979.

147. Lee, D., Castro, C. E., Jagodzinski, L. L., and Sevall, J. S., Functional significance of rat liver nonhistone protein-DNA interactions: RNA hybridization of protein-bound DNA, *Biochemistry,* 18, 3160, 1979.

148. Sevall, J. S., Jagodzinski, L. L., Tsai, S., Castro, C. E., and Lee, D., Nonhistone proteins and gene organization, in *The Cell Nucleus,* Vol. 4, Bush, H., Ed., Academic Press, New York, 1978, 319.

149. Gates, D. M. and Bekhor, I., DNA-binding assay of tightly-bound nonhistone chromosomal proteins in chicken liver chromatin, *Nuc. Acid. Res.,* 6, 3411, 1979.

150. Bekhor, I. and Mirell, C. J., Simple isolation of DNA hydrophobically complexed with presumed gene regulatory proteins, *Biochemistry,* 18, 609, 1979.

151. Berezney, R. and Coffey, D. S., Nuclear matrix, *J. Cell Biol.*, 73, 616, 1977.
152. Norman, G. L. and Bekhor, I., Distribution of poly (A) mRNA gene sequences and nonhistone chromosomal proteins in fractionated human chromatin, *J. Cell Biol.*, 87, CH45a, 1980.
153. Neubort, S., Liebeskind, D., Mendez, F., and Bases, R., Tightly bound chromatin proteins with rapid metabolic turnover suppress antinucleoside immunoreactivity in HeLa cell nuclei, *Exp. Cell. Res.*, 123, 55, 1979.
154. Pederson, T. and Bhorjee, J. S., A special class of non-histone protein tightly complexed with template-inactive DNA in chromatin, *Biochemistry*, 14, 3238, 1975.
155. Mayfield, J. E., Serunian, L. A., Silver, L. M., and Elgin, S. C. R., A protein released by DNase I digestion of *Drosophilia* nuclei is preferentially associated with puffs, *Cell*, 14, 539, 1978.
156. Hsieh, T. and Bruttag, D. L., A protein that preferentially binds *Drosophilia* satillite DNA, *Proc. Natl. Acad. Sci. U.S.A.*, 76, 726, 1979.
157. Galas, D. J. and Schmitz, A., DNAase footprinting: a simple method for the detection of protein-DNA binding specificity, *Nuc. Acid. Res.*, 5, 3157, 1978.
158. Bonne, C., Duguet, M., and de Recondo, A. M., Single-strand DNA binding protein from rat liver interactions with supercoiled DNA, *Nuc. Acid. Res.*, 8, 9955, 1980.
159. Cozzarelli, N. R., DNA gyrase and the supercoiling of DNA, *Science*, 207, 953, 1980.
160. Brown, P. O. and Cozzarelli, N. R., A sign inversion mechanism for enzymatic supercoiling of DNA, *Science*, 206, 1081, 1979.
161. Fareed, G. C. and Davoli, D., Molecular biology of papovaviruses, *Annu. Rev. Biochem.*, 46, 471, 1977.
162. Bonne, C., Duguet, M., and de Recondo, A. M., Rat liver DNA binding proteins: physiological variations, *FEBS Lett.*, 106, 292, 1979.
163. Weideli, H., Brack, C., and Gehring, W. J., Characterization of *Drosophilia* DNA-binding protein DB-2: demonstration of its sequence-specific interaction with DNA, *Proc. Natl. Acad. Sci. U.S.A.*, 77, 3773, 1980.
164. Sargent, T. D., Wu, J. R., Sala-Trepat, J. M., Wallace, R. B., Reyes, A. A., and Bonner, J., The rat serum albumin gene: analysis of cloned sequences, *Proc. Natl. Acad. Sci. U.S.A.*, 76, 3256, 1979.
165. Southern, E. M., Detection of specific sequences among DNA fragments separated by gel electrophoresis, *J. Mol. Biol.*, 98, 503, 1975.
166. Tijian, R., The binding site on SV 40 DNA for a T-antigen related protein, *Cell*, 13, 165, 1978.
167. Buckler-White, A. J., Humphrey, G. W., and Pigiet, V., Association of polyoma T-antigen and DNA with the nuclear matrix from lytically infected 3T6 cells, *Cell*, 22, 37, 1980.
168. Franke, B. and Eckard, W., Polyoma gene function required for viral DNA synthesis, *Virology*, 55, 127, 1973.
169. Martin, R. G. and Oppenheim, A., Initiation points for DNA replication in nontransformed and Simian Virus 40-transformed Chinese hamster lung cells, *Cell*, 11, 859, 1972.
170. Giacherio, D. and Hager, L. P., A specific DNA unwinding activity associated with SV 40 large T-antigen, *J. Biol. Chem.*, 255, 8963, 1980.
171. Pardoll, D. M., Vogelstein, B., and Coffey, D. S., A fixed site of DNA replication in eucaryotic cells, *Cell*, 19, 527, 1980.
172. Vogelstein, B., Pardoll, D. M., and Coffey, D. S., Supercoiled loops and eucaryotic DNA replication, *Cell*, 22, 79, 1980.
173. Barrack, E. R. and Coffey, D. S., The specific binding of estrogens and androgens to the nuclear matrix of sex hormone responsive tissue, *J. Biol. Chem.*, 255, 7265, 1980.
174. Thrall, C. L. and Spelsberg, T. C., Factor affecting the binding of chick oviduct progesterone receptor to deoxyribonucleic acid: evidence that deoxyribonucleic acid alone is not the nuclear receptor site, *Biochemistry*, 19, 4130, 1980.
175. Roeder, R. G., Eucaryotic nuclear RNA polymerases, in *RNA Polymerase*, Losich, R. and Chamberlain, M., Eds., Cold Spring Harbor Laboratory, Cold Spring Harbor, N.Y., 1976, 285.
176. Brown, D. D., Wensink, P. C., and Jordan, E., Purification and some characteristics of 5S DNA from *Xenopus laevis*, *Proc. Natl. Acad. Sci. U.S.A.*, 68, 3175, 1971.
177. Brown, D. D. and Gurdon, J. B., Cloned single repeating units of 5S DNA direct accurate transcription of 5S RNA when injected into *Xenopus* oocytes, *Proc. Natl. Acad. Sci. U.S.A.*, 75, 2849, 1978.
178. Brown, D. D. and Gurdon, J. B., High-fidelity transcription of 5S DNA injected into *Xenopus* oocytes, *Proc. Natl. Acad. Sci. U.S.A.*, 74, 2064, 1977.
179. Wu, G. J., Adenovirus DNA-directed transcription of 5.5S RNA *in vitro*, *Proc. Natl. Acad. Sci. U.S.A.*, 75, 2175, 1978.
180. Birkenmeier, E. G., Brown, D. D., and Jordan, E., A nuclear extract of *Xenopus laevis* that accurately transcribes 5 RNA genes, *Cell*, 15, 1077, 1978.
181. Weil, P. A., Segall, J., Harris, B., Ng, S. Y., and Roeder, R. G., Faithful transcription of eucaryotic genes by RNA polymerase III in systems reconstituted with purified templates, *J. Biol. Chem.*, 254, 6163, 1979.

182. Sakonju, S., Bogenhagen, D. F., and Brown, D. D., A control region in the center of the 5S RNA gene directs specific initiation of transcription. I. The 5′ border of the region, *Cell*, 19, 13, 1980.
183. Bogenhagen, D. F., Sakonju, S., and Brown, D. D., A control region in the center of the 5S RNA gene directs specific initiation of transcription. II. The 3′ border of the region, *Cell*, 19, 27, 1980.
184. Engelke, D. R., Ng, S. Y., Shastry, B. S., and Roeder, R. G., Specific interaction of a purified transcription factor with an internal control region of 5S RNA genes, *Cell*, 19, 717, 1980.
185. Pelham, H. R. and Brown, D. D., A specific transcription factor that can bind either the 5S RNA gene or 5S RNA, *Proc. Natl. Acad. Sci. U.S.A.*, 77, 4170, 1980.
186. Honda, B. M. and Roeder, R. G., Association of a 5S gene transcription facor with 5S RNA and altered levels of the factor during cell differentiation, *Cell*, 22, 119, 1980.
187. Bittner, M., Kupferer, P., and Morris, C. F., Electrophoretic transfer of proteins and nucleic acids from slab gels to diazobenzyloxymethyl cellulose sheets, *Anal. Biochem.*, 102, 459, 1980.
188. Hipskind, R. A. and Reeder, R. H., Initiation of ribosomal RNA chains in homogenates of oocyte nuclei, *J. Biol. Chem.*, 255, 7896, 1980.
189. Wasylyk, B., Kidinger, D., Corden, J., Brison, O., and Chambon, P., Specific *in vitro* initiation of transcription on conalbumin and ovalbumin genes and comparison with adenovirus-2 early and late genes, *Nature (London)*, 285, 367, 1980.
190. Jelinek, W. R., Toomey, T. P., Leinwand, L., Duncan, C. H., Biro, P. A., Chouday, P. V., Weissman, S. M., Rubin, C. M., Houck, C. M., Deininger, P. L., and Schmid, C. W., Ubiquitous, interspersed repeated sequences in mammalian genome, *Proc. Natl. Acad. Sci. U.S.A.*, 77, 1398, 1980.
191. Krayev, A. S., Kramerov, D. A., Skryabin, K. G., Ryskov, A. P., Bayev, A. A., and Georgiev, G. P., The nucleotide sequence of the ubiquitous repetitive DNA sequence B1 complementary to the most abundant class of mouse fold-back RNA, *Nuc. Acid. Res.*, 8, 1201, 1980.
192. Lapeyre, J. N. and Becker, F. F., Analysis of highly repeated DNA sequences of rat with Eco RI endonuclease, *Biochim. Biophys. Acta*, 607, 23, 1980.
193. Duncan, C., Biro, P. A., Chouday, P. V., Elder, J. T., Wang, R. R. C., Forget, B. G., Riel, J. K., and Weissan, S. M., RNA polymerase III transcriptional units are interspersed among human non-α-globin genes, *Proc. Natl. Acad. Sci. U.S.A.*, 76, 5095, 1979.
194. Fritsch, E. F., Lawn, R. M., and Maniatis, T., Molecular cloning and characterization of the human β-like globin gene cluster, *Cell*, 19, 959, 1980.
195. Jagodzinski, L. L., personal communication.

Chapter 3

ROLE OF NONHISTONE CHROMATIN PROTEINS AND SPECIFIC DNA SEQUENCES IN THE NUCLEAR BINDING OF STEROID RECEPTORS

Thomas C. Spelsberg

TABLE OF CONTENTS

I. INTRODUCTION

The nonhistone proteins of chromatin have been suspected to play a role in the regulation of gene expression in the chromatin of eukaryote cells. The biological function of a single nonhistone protein, besides enzymatic, has yet to be definitively defined. One approach to the study of the cellular components involved in gene regulation is to employ extracellular agents which alter patterns of gene expression. Steroid hormones are naturally occurring agents which regulate the fate and function of cells via regulating patterns of gene transcription.

Steroids enter cells from the vascular system and bind to specific cellular proteins called receptors. These receptors are found only in "target" cells which respond to the steroid. Each class of steroids binds to its own receptor with high affinity. This binding to the receptor results in the translocation and binding of the steroid receptor complex to the nuclear binding sites or nuclear "acceptor" sites as first termed for these nuclear sites by Spelsberg et al.[1] The nuclear binding in turn results in immediate alterations in the pattern of transcription. These events occur within 1 hr after the entrance of the steroid into the cell. Thus, it is likely that steroid receptors may represent the first class of intracellular gene regulators ever discovered in eukaryote cells.

They are triggered by an external effector (steroids) to bind to nuclear acceptor sites and alter gene expression and then return to the cytoplasm for the next effector. It seems inherently important that the focal point of the action of steroid receptor complexes on chromatin must comprise components involved in the regulation of gene transcription.

This laboratory has been investigating the nuclear interactions of steroid receptors for the past decade. This chapter will briefly describe the chemical properties of the chromatin acceptor sites obtained over the past 6 years for the progesterone receptor in the avian oviduct. These nuclear binding sites appear to involve a specific class of nonhistone proteins and specific sequences of DNA which is the subject of this chapter. The DNA sequences are associated with repetitive sequences. Interestingly, another class of nonhistone proteins appears to be involved in masking many of the chromatin acceptor sites. The implications of this masking will also be discussed in this chapter. Before describing the properties of nuclear acceptor sites for steroid receptors, a moderately detailed description of the mechanistic pathway for steroid hormones in target cells is presented.

II. SPECIFICS ON THE MECHANISM OF ACTION OF STEROID HORMONES

The same general theme for the mechanism of action of steroids has been identified in all steroid-target tissue systems in all animals. This is outlined in Figure 1. For more comprehensive reviews, the readers are referred elsewhere.[2-6] Briefly, systemic steroids, are found to be mostly bound to serum carrier proteins. Those that are unbound enter target cells. Whether the transfer involves active transport or passive diffusion remains in question.[7-10] Whether a steroid enters nontarget tissues is also open to question. Radiolabeled steroids are found in the cytosols of nontarget cells;[11,12] however, these results may be explained by an artificial transfer of the steroids from the blood compartments of organs to the "cytosol" during tissue homogenization. In any event, the steroids are retained in their target cells but not in nontarget cells for extended periods.[11-13] This retention is due to specific steroid binding proteins termed receptors which are found only in target cells.[14-18] Although the exact intracellular compartment in which unbound receptors reside remains obscure, there is some evidence that they are associated with microsomes[19,20] while others claim they might reside in the nucleus.[21,22]

Once bound by the steroid, the receptor becomes "activated", a process which invests in the complex the capacity to "migrate" into the nucleus and bind to nuclear acceptor sites. These processes of activation and translocation of a steroid receptor complex into the nucleus are not well understood. For instance, exactly from what site the migration occurs is unknown. Cytosolic factors exist which can induce (enhance) activation[24-27] and others which suppress activation or translocation.[28-34] Still others have presented evidence that the native form of at least one receptor (the chick oviduct progesterone receptor) is composed of two molecular species.[35-39] It is speculated that the two species act as subunits to a dimer. The latter represents the active form of the receptor.[38-42] It is further speculated that this dimer then migrates to the nucleus where the two different subunits (termed A and B) perform distinct functions. The B subunit binds the complex to specific acceptor sites on the chromatin, while A alters transcription by binding the adjacent DNA sequence. This active subunit model, however, requires further study.

Once bound to nuclear acceptor sites, the steroid-receptor complex markedly alters RNA synthesis[43-45] and the transcription of specific genes.[46-50] The subsequent effects of steroids on RNA and protein processing, if any, are poorly understood due to the few studies undertaken.[51]

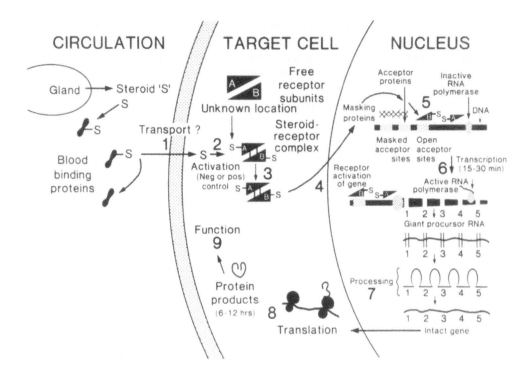

FIGURE 1.   Basic pathway for the action of steroid hormones in target cells. This pathway is described in the text. It represents the general mechanism of action of all steroids. Briefly, beginning on the left side of the model, the steroid (H) passes through the cell membrane and binds to its specific receptor. The receptor, represented as two subunits, is then "activated" and the complex of steroid-receptor translocates and binds to the nuclear acceptor sites. This results in changes (quantitative and qualitative) in DNA-dependent RNA synthesis. The giant precursor mRNA molecules are "processed" to small mRNA molecules which are transported out of the nucleus to the ribosomes where they code for specific steroid-induced proteins. These proteins carry out the various steroid-induced physiological responses of the target cell. In essence, the first major intracellular process which responds to the steroids is the alteration in gene transcription (RNA synthesis). This is preceded by the binding of the steroid receptor to the nuclear acceptor sites. (Data for this schema were modified from Thrall, C. L., Webster, R. A., and Spelsberg, T. C., *The Cell Nucleus*, Vol. 6, Part 1, Busch, H., Ed., Academic Press, New York, 1978, 461; and from O'Malley, B. W. and Means, A. R., *Science,* 183, 610, 1974).

On a chronological basis, the above events occur rapidly. Within 1 to 2 min after injection of labeled progesterone into chicks, which have been primed with estrogen, the steroid is located in the cytosol bound to its receptor. By 10 min postinjection, the steroid accumulates in the nucleus, followed closely by changes in RNA synthesis and chromatin template capacity. After 1 to 2 hr specific messenger RNAs appear whose coded proteins begin to accumulate within 3 to 6 hr. If only a single injection of the steroid is used, the steroid-induced changes rapidly return to control values. At 12 hr postinjection, the radiolabeled steroids in the cytoplasm and nucleus are markedly reduced as are the alterations in RNA synthesis. The fate of the steroids is unclear but it is probable they are not reutilized but are metabolized and excreted from the cells. Whether or not the receptor is reutilized to some degree in the steroid action pathway is unknown. At least some of the receptor is thought to be recycled.[52,53]

Thus, the binding of the steroid-receptor complex to acceptor sites on the chromatin represents the first nuclear event which occurs before the steroid-induced changes in transcription occur. The receptor when bound by a specific steroid is triggered to migrate into the nucleus, bind to chromatin acceptor sites, and alter transcription. The whole process is rapidly reversible to allow immediate control of the target tissue by the steroid. It is the author's opinion that the steroid receptors may well represent one of the first, if not the first, intracellular gene regulators discovered in eukaryotes.

III. CHEMICAL PROPERTIES OF NUCLEAR ACCEPTOR SITES FOR STEROID RECEPTORS

A. General

The exact biological components comprising the nuclear acceptor sites for steroid receptors remains elusive. The following have been implicated as being involved in the nuclear acceptor sites for various steroid receptors: the nuclear envelope,[54,55] histones,[56-59] nonhistone basic proteins,[60-62] nonhistones and nonhistone-DNA complexes,[1,6,37,57,58,63-83] pure DNA,[30,32,72,73,84-96] ribonucleoproteins,[97] and finally the nuclear matrix.[98,99]

Of the many components suggested, the DNA and DNA-nonhistone protein complexes have received the most attention. A few studies of the nuclear components whose role in nuclear acceptor sites is still regarded as potential are described in more detail in the following sections.

B. Chromatin Protein as Acceptor Sites

Puca et al.[60,61] and Mainwaring et al.[62] described a basic protein with a molecular weight of 60,000 which appears to have high affinity for the estrogen receptor in the calf uterus and for the androgen receptor in rat prostate, respectively. The binding occurs in the absence of DNA. Since other nonchromatin basic proteins demonstrate similar estrogen receptor binding[60,61] and since no further reports for these acceptors have been published in the past 6 years, these will not be discussed further. There have been two recent reports on the localization of a nuclear acceptor for the estrogen receptor in the nuclear matrix. Since this matrix contains minimal amounts of DNA, these nuclear binding sites may involve only proteins. The nuclear matrix studies are discussed below in Section III.E.

C. DNA as Acceptor Sites

It is undisputed that in cell-free conditions steroid receptors bind to DNA. Many laboratories have reported that a variety of steroid-receptor complexes bind to DNA.[1,26,30,32,66-70,72,73,81,85,87-91,96,100-107]

The binding of the receptor to DNA appears to be in part electrostatic interaction, decreasing with increasing ionic strength.[2,3,73,89,96,108-111] The same is true of receptor binding to nuclei and chromatin.[81,112] However, steroid-receptors are more readily dissociated from the DNA than from chromatin or nuclei at higher ionic strengths.[111] The interaction of steroid receptors with pure DNA apparently is not totally electrostatic in nature as evidenced by the inhibitory effect of the intercalating drugs, ethidium bromide and 9-hydroxyellipticine, on receptor binding to DNA.[113] These compounds significantly reduce receptor binding to double-stranded DNA-cellulose, but not receptor binding to phosphocellulose or polyadenylic acid-cellulose. This suggests an interaction of receptors with the bases of DNA involving some hydrophobic interactions.

The exact physiological significance of steroid receptor-DNA interactions remains to be demonstrated. Several lines of evidence support that DNA might play a role in the acceptor sites for steroid receptors. First, Yamamoto et al.[110] reported that the response of mutant clones of cultured mouse lymphoma cells to glucocorticoids is paralleled by the DNA affinity of the receptor in vitro. Second, there have been reports of pseudo-specific interactions in terms of preferences of steroid receptors for substituted over unsubstituted DNAs,[106] for native over denatured DNAs,[103,113] for poly(dA-

T),[94,103] oligo(dT) or oligo(dC)[26,105] over other synthetic polydeoxyribonucleotides, for eukaryotic over prokaryotic DNA,[85] and finally for native DNA over RNA.[87,89,96] Lastly, DNase treatment of nuclei appears to reduce receptor binding or release previously bound receptor in a variety of steroid-target tissue systems.[72,73,84,92-94,114] This supports the notion that DNA might play a role in acceptor activity.

In contrast, however, many facts point to a nonspecific interaction of steroid receptors with pure DNA. The affinity of the progesterone receptor for the DNA of the chick oviduct is significantly lower than that of the receptor for nuclei and chromatin.[115] Furthermore, the differences in binding of the steroid receptors to different DNA sequences are relatively small. In fact, many laboratories have found no differences in the binding of steroid-receptor complexes to the native DNAs of a variety of sources.[89,98,100,102,103] In most instances, the binding to pure DNA is not saturable.[30,32,81,89,94,96,100,105,111,115] It is possible that the lack of DNA specificity for steroid receptors could be due to the inability to detect a few specific sites among an overwhelming number of slightly lower affinity, nonspecific sites.[90,91] Proof for these few specific sites, however, has yet to be reported.

Recent studies by Thrall and Spelsberg[107] indicate that the binding of the chick oviduct progesterone receptor to DNA does not correlate with the native (in vitro) binding. Further, these authors identified specific artifacts in the DNA-binding assays. The chick oviduct progesterone receptor displays a variable, nonsaturable binding to pure DNA in cell-free binding assays whereas a saturable binding is observed with whole chromatin or a protein-DNA complex. Three factors are identified which affect the binding of P-R to the DNA. These are (1) the conditions of the binding assay, (2) the particular receptor preparation, and (3) the state of the DNA.[107] The conditions in the binding assay which affect the extent of DNA binding are the choice of the blanks, the salt concentration, and the pH of the assay. Concerning the second class of factors controlling DNA binding, various preparations of the receptor display their own characteristic levels of binding to native DNA, the basis of which is unknown. Lastly, the purity and the integrity of the DNA also determines the level of binding of the P-R. Protein impurities as well as moderate degradation of the DNA by enzymatic or physical fragmentation and ultraviolet (UV) light treatment greatly enhance the receptor binding to the DNA. Interestingly, totally denatured (single-stranded) DNA displays little or no binding of the P-R.[107] Thus, the extent of binding of the steroid receptor depends on the degree of damage to the DNA.

As will be shown later, seasonal differences which are observed for the binding of P-R to chromatin in vivo and in vitro do not occur with pure DNA. Thrall and Spelsberg[107] concluded from their studies that under controlled conditions and by using DNA preparations as undamaged as possible, minimal binding of P-R to pure DNA occurs. They further concluded that native or partially degraded, pure DNA alone does not appear to represent the native nuclear acceptor sites for the P-R in the chick oviduct.[107] In contrast, the DNA-nonhistone protein (acceptor protein) complexes do show characteristics of the native-like acceptor sites. It is the authors' opinion that the numerous reports in the literature describing marked binding of the steroid-receptor complex to DNA are probably due to one of the conditions listed above and that DNA alone does not have the properties characteristic of the binding in vivo.

D. DNA-Protein Complexes as Acceptor Sites

As mentioned earlier, many laboratories have reported evidence supporting a combination of DNA and protein as a complex in chromatin serve as the nuclear acceptor sites for a variety of steroid receptors.[1,6,38,39,60,61,64,65,68-82] Most of these nucleoprotein complexes contain specific fractions of nonhistone chromatin proteins bound to DNA and display high affinity, saturable nuclear binding sites for steroid receptors (i.e., acceptor activity).

Several lines of evidence support these nonhistone protein-DNA complexes (nucleoacidic protein or NAP) as the nuclear acceptor sites for the progesterone receptor in the avian oviduct.[1,6,39,64,65,68,69,111] First, the nonhistone proteins which when removed from the chromatin DNA result in a loss in P-R binding can be reannealed to pure DNA to yield a reconstituted "NAP" which contains the same number of acceptor sites as found in native (undissociated) NAP.[6,39,67,70,116] Reconstitution, using other protein fractions from the chromatin, fail to reconstitute P-R binding levels. These data are further discussed in later sections. In any case, the acceptor activity appears to be specific for certain proteins. Second, recent studies have indicated that the specific acceptor proteins require specific DNA sequences to achieve acceptor activity, i.e., binding the progesterone receptor. When the proteins are reannealed to the DNA of different biological sources, only the reconstituted NAP containing the hen DNA binds the progesterone receptor.[200] This aspect is further discussed in a later section. Therefore, the acceptor activity appears to be specific for certain sequences of the DNA. The third reason to suspect that the acceptor protein-DNA complexes represent the native acceptor sites is the fact that the binding of P-R to native chromatin or NAP requires an intact receptor. It has been shown elsewhere that the cell-free binding to the chromatin or NAP also requires an "activated" receptor[81] which is a characteristic of the in vivo binding.[3] The fourth reason comes from correlations between in vivo and in vitro nuclear binding of P-R.[39,107,117] In these studies the pattern of in vitro binding in assays using chromatin or NAP were shown to correlate closely with those in vivo. The binding assays using pure DNA or the DNA bound with other protein fractions, however, showed no such correlation.

E. The Nuclear Matrix as Acceptor Sites

Barrack and co-workers[98,99] have reported that as much as 67% of the total nuclear bound dihydrotestosterone binding sites with properties of the receptor and 100% of the salt resistant nuclear receptors are localized in the nuclear matrix of the rat ventral prostate. Similar results were obtained with estrogen in the chicken liver. Structurally this nuclear matrix is composed of a residual nuclear envelope with pore complexes, remnants of an internal fibrogranular network, and the residual nucleoli. Chemically, it contains 7% of the total nuclear protein and 2% of the total cellular DNA. They further found that a subfraction of the nuclear matrix, the internal ribonucleoprotein network, involving the internal network but not the peripheral lamina, contained the majority of the androgen binding sites associated with the whole matrix.[95] These sites appear to be destroyed by a combination of DNase treatments, RNase, and dithiothreitol. Interestingly, these sites displayed expected quantitative changes under different endocrine states of the animals. Unfortunately, these analyses of the steroid "receptors" bound to the nuclear matrix involve an indirect binding assay, i.e., steroid exchange. As mentioned by these authors, the chemical nature of these binding sites remains obscure. These sites (receptors) are not extracted by $2M$ NaCl extractions in contrast to the bulk of the natively bound nuclear steroid receptor which is extracted. These binding sites for the steroids, however, do display many properties characteristic of steroid receptors including saturability, steroid specificity, high affinity, and low capacity.

The exact chemical characterization of the nuclear acceptor sites identified with the nuclear matrix awaits further investigation. It is interesting that the sites appear to be destroyed by a combination of DNase and RNase treatments or by dithiothreitol. It is possible that the acceptor sites in the nuclear matrix represent protein-DNA complexes as discussed for the avian oviduct progesterone receptor. The small amounts of DNA and protein located in the nuclear matrix could well contain the acceptor sites as described above for the oviduct progesterone receptor.

IV. PROGRESS AND PROBLEMS IN THE CHEMICAL
CHARACTERIZATION OF ACCEPTOR SITES

Studies on the intranuclear localization of steroid-receptor complexes for estrogen,[3,63,118,119] progesterone,[1,6,65-70,76,81,116,120,121] dihydrotestosterone,[77] glucocorticoids,[83,93] and aldosterone have been numerous.[121]

Identifying the component of chromatin which binds the steroid-receptor complex (acceptor sites) has been the object of much attention in the past 12 years. The history of these studies has shown that: (1) proper cell-free nuclear binding assays must be established, (2) analysis should be performed to verify that the nuclear sites being measured are similar to those measured in vivo, (3) the extent of contamination of the chromatin, e.g., nuclear envelope, nuclear matrix, cytoplasmic entities, etc., should be measured, and (4) the approach and methods to identify the chemical nature of the acceptor sites should be carefully chosen.

A. Problems Confronted in the Specific Enzyme Treatments of Chromatin

There are many reports which claimed that the nuclear acceptors must be proteins since the steroids are released by protease treatment of chromatin.[59,121-123] However, the proteases are very likely degrading the receptor itself, thus releasing the steroid. Similarly, studies using deoxyribonuclease treatment of chromatin are difficult to interpret. First there is controversy as to whether DNA is the sole acceptor site in that same claim that the acceptor sites are destroyed by DNase[72,73,124,125] while others claim that the acceptor sites are not destroyed by DNase.[86,114] Second, if the nuclease does release bound steroid, it does not prove that DNA is the sole acceptor site since the DNA may represent only a part of the acceptor site (e.g., a co-acceptor) or the nuclease may cause the release of intact "acceptor" sites (e.g., protein or protein-DNA complexes) from the chromatin. As a result of these problems, most investigators began studying the cell-free nuclear binding of steroid-receptor complexes to chromatin acceptor sites.

B. Development of a Cell-Free Nuclear Binding for the Chick Oviduct Progesterone Receptor

The basic procedure for performing cell-free nuclear binding studies is to incubate a labeled steroid receptor preparation with isolated nuclei or nuclear component, and then measure bound vs. free steroid receptor. The numerous reports involving such studies utilize some variation of this approach. These variations include the level of purification of the receptor, the method of "activation" of the steroid-receptor complex (e.g., salt treatment, dilution, partial purification, or heat activation, etc.), the experimental design used in analyzing nuclear uptake (e.g., varying the receptor levels, varying the nuclear levels, etc.), the conditions of the cell-free assay (e.g., temperature of incubation, ionic strength, etc.), and the method of removing unbound or nonspecifically bound steroid-receptor complex from the nuclear components.

Using these considerations, a cell-free nuclear binding assay for the progesterone receptor in the chick oviduct was developed. This is outlined in Figure 2 for both whole nuclei/chromatin as well as for various soluble nuclear components which require special handling. Since the free steroid-receptor complexes are soluble but the nuclei and chromatin are not, it is easy to separate the chromatin bound from unbound steroid-receptors by using short-term centrifugation. The partially deproteinized chromatins, however, are soluble and thus are more difficult to separate from the soluble receptors. Long periods of centrifugation were initially used to separate these two soluble entities, but low recovery, degradation, and limited sample numbers presented problems.[1,64] A better approach is to rapidly render the soluble nucleoproteins insolu-

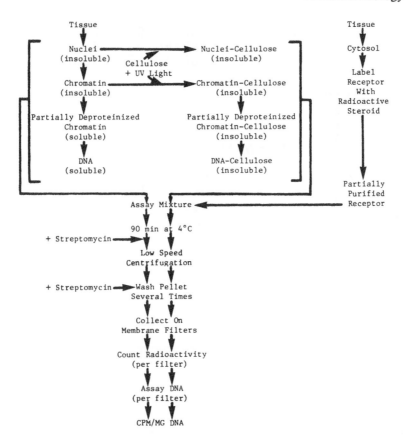

FIGURE 2.   Outline of basic procedure for the cell-free assay of [³H]P-R binding to nuclear acceptor sites. These methods are described in detail elsewhere.[67-70,81,116,117,127] The methods were selected on the basis of achieving saturable binding to the highest affinity class of sites and for maintaining the integrity of the steroid-receptor complex, the chromatin protein, and its DNA (see text). (From Thrall, C. L., Webster, R. A., and Spelsberg, T. C., *The Cell Nucleus*, Vol. 6, Busch, H., Ed., Academic Press, New York, 1978, 461. With permission.)

ble without damaging the steroid receptor or its binding to the acceptor sites. This is accomplished in either one of two ways as shown in Figure 2. First, the chromatin can be attached to an insoluble resin such as cellulose or acrylamide and the various protein fractions removed, leaving residual DNA or DNA-protein complex attached to the resins in an insoluble state. These resin can then be used in the binding assays. Alternately, one can perform the binding using free nucleoproteins after the hormone binding, the soluble chromatin fractions can be rendered insoluble using streptomycin sulfate to precipitate the nucleoprotein with bound steroid receptors. In each method, the complexes of DNA and steroid receptor are washed several times to remove traces of the unbound steroid receptor. The washed complexes are then transferred to millipore filters and the bound radioactivity counted. The DNA per filter is quantitated and the radioactivity per milligram DNA or molecules of bound P-R per cell are calculated as described elsewhere.[81,112] These methods as well as their problems and limitations are detailed elsewhere.[6]

The assay conditions utilize a partially purified receptor (20-fold purification) which is preactivated,[112,126] a low temperature (4°C), a 90 min incubation period, and varying ratios of receptor/nuclear DNA with 30 to 100 μg DNA (as nuclei or nuclear components) per assays. The time of incubation was selected to allow equilibrium between bound (to acceptor sites) and free progesterone receptor in the range of receptor and

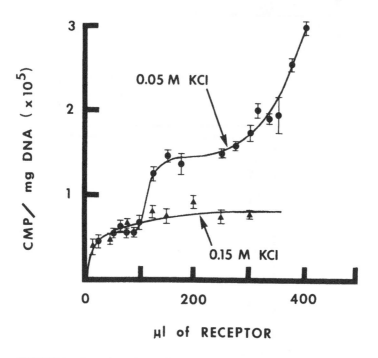

FIGURE 3.   Detection of secondary classes of nuclear binding sites in vitro in the hen oviduct; effects of ionic strength. These experiments were performed with the oviducts from mature hens as described elsewhere.[111,112] The assay mixtures contained either (●) 0.05 M KCl or (▲) 0.15 M KCl (final concentration). The assays were performed using 25 μg of DNA per assay and the standard assay method. The range and average of three replicates of analyses for each receptor concentration are shown. (From Pikler, G. M., Webster, R. A., and Spelsberg, T. C., *Biochem. J.*, 156, 399, 1976. With permission.)

chromatin concentrations used. As shown in Figure 3, these conditions allow a saturable binding in a short period of time using increasing amounts of the receptor with constant amounts of nuclear components. Figure 4 demonstrates that saturation is also observed when the nuclear components are varied with constant amounts of receptor. The affinity of this cell-free nuclear binding has been determined by Scatchard analysis with a K_d of $10^{-9} M$.[81,112] We have found that 90% of the progesterone receptor remains intact throughout the assay,[6,112] the chromatin proteins and DNA show relatively little damage,[127] and about 60 to 80% of the chromatin DNA is recovered on the filters after the binding, washings, and transfer. Further, when [^{14}C]ovalbumin is added to the binding assays and then assayed by SDS polyacrylamide gel electrophoresis, less than 10% of the ovalbumin is damaged.[201] In contrast, when the cell-free binding assays are performed under conditions of higher temperature (25°C), for longer periods or with crude cytosol, significant proteolytic damage to the [^{14}C]ovalbumin is observed. This proteolytic activity is accompanied by increases in binding of the progesterone receptor to nuclear acceptor sites, as shown in Figure 5.[202] The use of mild protease inhibitors such as phenylmethylsulfonyl fluoride offer additional protection to the proteolytic activity during the binding assays. Potent protease inhibitors may damage the receptors and prevent nuclear binding.

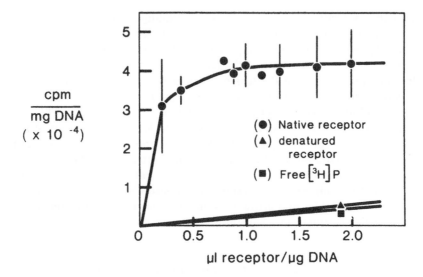

FIGURE 4.   Titration of P-R binding to hen oviduct chromatin. The cell-free nuclear binding of (●) native P-R, (▲) P-R denatured by heating at 50°C for 1 hr, and (■) free steroid to oviduct chromatin, were performed as described elsewhere[112] except that the levels of chromatin were varied and the P-R concentration per assay kept constant. The mean and standard deviation of four replicate binding assays to the native receptor are presented.[201]

SUSCEPTIBILITY OF MASKING ACTIVITY AND ACCEPTOR ACTIVITY TO PRONASE

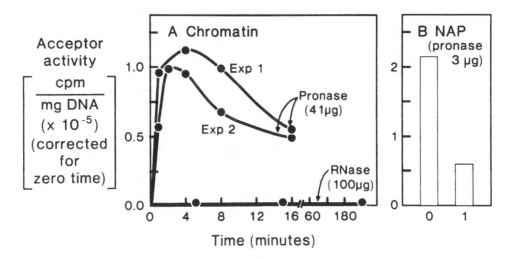

FIGURE 5.   Effects of proteolytic action on oviduct chromatin and its consequences on the binding of the oviduct progesterone receptor (P-R). Oviduct chromatin (Panel A) and oviduct nucleoacidic protein (NAP) (Panel B) were treated with pronase at a ratio of pronase to DNA (w/w) of 0.05 at 4°C in dilute Tris buffer (pH 7.5). At various intervals, the aliquots were removed, diluted tenfold with buffer, and centrifuged at $10^5 \times g$ for 12 hr. The DNA-residual protein pellets were resuspended in dilute buffer and assayed for acceptor activity (P-R binding). In Panel A (●) P-R binding to the pronase- or ribonuclease-treated chromatins is indicated. The ribonuclease/DNA (w/w) was 0.2. The mean of four replicate binding assays are presented for each value. In Panel B the bar graph indicates the P-R binding.[202]

C. Some Factors Affecting the Cell-Free Nuclear Binding Assays

Many problems confront the establishment of cell-free nuclear binding assays. The cell-free binding assays eliminate many of the obstacles that the in vivo assays faced such as need of expensive amounts of the radiolabeled steroid, difficulty in saturating the nuclear acceptor sites, and the loss of nuclear bound steroid-receptor complexes during isolation and manipulation of the chromatin. The latter is expected due to the noncovalent binding of the steroid-receptor complex to nuclear sites. This rapid dissociation prevents the fractionation of chromatin to identify the chemical nature of the nuclear acceptor sites for a particular steroid receptor. The cell-free binding approach on the other hand allows the chemical characterization of the acceptor sites by the selective dissociation of certain chromatin components followed by the binding of the steroid-receptor complex to the residual nucleoprotein. The cell-free assay, however, does include variables and artifacts, some intrinsic to the assay and some to the receptor and nuclei/chromatin preparations. These are mentioned in the following sections.

1. Protease and Nuclease Activities in the Chromatin and Receptor Preparations

Both chromatin[127-132] and the receptor preparations (especially, the crude cytosols)[42,127,133-136] contain nuclease and protease activities. The proteases can degrade the steroid-receptor complexes and prevent binding to nuclear acceptor sites.[133,134,136] Partial purification of the avian oviduct progesterone receptor removes most, if not all, of the protease activity, resulting in a better maintenance of the integrity of the receptor and chromatin during binding assays.[42,133-138]

Figure 5 shows that mild protease action on oviduct chromatin will cause an increase in the number of chromatin acceptor sites for the progesterone-receptor complex (P-R), whereas severe protease action will cause a marked decrease in the number of acceptor sites. The increase in acceptor sites with mild protease action is due to the destruction of masking proteins which cover many acceptor sites in target tissues and all acceptor sites in nontarget tissues.[6,66-69] This masking is further discussed later. If measures are not taken to nearly eliminate protease activity, increased binding to both target and nontarget tissue chromatins will occur, producing an apparent lack of saturable binding to nuclear sites[29,139] as well as a loss of tissue specific nuclear binding.[29,92-94,140,141] If extensive proteolysis is permitted, then a loss of nuclear binding due to the degradation of the acceptor protein is observed. More details of this proteolysis are discussed later. Thus, assay conditions should be established to reduce protease action by using low temperatures (4°C), previously activated receptors, partially purified receptor preparations, high ionic strength, and as clean preparation of chromatin as possible.[6,127]

Nucleases have also been shown to degrade the chromatin DNA during cell-free nuclear binding assays under certain conditions.[127] The binding to isolated chromatin can be enhanced or decreased as a result of mild or extensive nuclease action, respectively. Therefore, the conditions for the cell-free binding assay should include those which also minimize the nuclease activity. Fortunately, many of these conditions are the same as those required to minimize protease activity.[127]

2. Use of Proper Ionic Conditions to Identify High Affinity Nuclear Acceptor Sites

It has been shown that the ionic condition in the cell-free binding assay has a marked influence on the saturability and level of binding of the P-R to oviduct nuclear acceptor sites.[81,111,112] Figure 3 shows the effects of using a low (0.5 M KCl) and a high (0.15 M KCl) salt concentration on the cell-free binding of labeled progesterone receptor to oviduct chromatin. At the higher salt concentrations, i.e., physiological levels or higher, only the high affinity class of nuclear binding sites is detectable for the chick oviduct progesterone receptor. Using lower ionic conditions, a nonspecific nonsaturable binding is observed.[111] Based on these studies, the ionic conditions of 0.15 to 0.18 M KCl are recommended for the cell-free binding assays. As discussed later, the ionic environment together with the degree of proteolytic activity explain in part the reported lack of saturable nuclear binding using cell-free assays in certain laboratories.

3. Variations in the Levels and Biological Activity of the Receptors

We have recently reported that the chick oviduct progesterone receptor varies at different times of the year both in quantity and in biological function, i.e., nuclear binding and alteration of transcription.[39,117] In the winter the progesterone receptor decreases to half the normal level and loses practically all of its biological activity. This winter period is accompanied by a loss in the nuclear binding of the chick oviduct progesterone and estrogen receptors using both in vivo and in vitro binding assays. Similar seasonal effects are seen in the estrogen receptor.[203] These results are further discussed later. Other studies have shown that the progesterone receptors from the immature oviduct, from estrogen withdrawn chicks, or from molting hens are also not capable of nuclear translocation and binding.[204] Thus, the progesterone receptors of the chick oviduct are not always constant in levels or biological activity, a factor which must be considered when performing cell-free assays. Whether such events occur in other animals remains to be determined.

As discussed earlier, the cytosol of many target tissues contains small and/or large molecules which appear to regulate the receptor activation and nuclear binding. The cytosols also contain proteases and nucleases as discussed previously. Partial purification of the receptors by ammonium sulfate precipitation appears to separate some of the regulatory factors as well as the proteases from steroid receptors.[25,42,136-139] Therefore, it is recommended that cell-free binding assays utilize partially purified receptor preparations to reduce or eliminate these factors and enzymes activities. Other studies have shown that the titration of crude steroid-receptor complexes with constant amounts of chromatin to determine saturation of nuclear acceptor sites may involve an artifact due to the variation in protein concentration in the assays. There are several ways to assess this possible artifact. First, the use of partially purified receptors may alleviate excess variations in protein levels and reduce the interference with acceptor site saturation analysis. Second, the cell-free assays can be organized to keep the cytosol protein constant and vary the nuclear acceptor sites. Lastly, the binding assays could be arranged so that the chromatin DNA is kept constant while the receptor levels vary. In this case, equivalent amounts of protein are maintained by adding albumin or the cytosol protein from nontarget tissues containing no receptor.

The studies discussed in this last section demonstrate that the physiological state of the animal may have a large impact on the reproducibility of cell-free binding assays and the estimation of the number of nuclear acceptor sites. To identify such variations of steroid receptors, extensive analysis of the nuclear binding of the steroid-receptor complex is required.

4. Problems with Certain Steroids

When both the estrogen and progesterone receptors are used for nuclear acceptor site analysis, a marked difference in the adsorption between the two steroids to nuclei and nuclear components is observed.[142] In our hands, free estrogen displays as much as 5- to 10-fold greater nonspecific adsorption to nuclear fractions than does free progesterone. In order to demonstrate a receptor dependent, specific binding to nuclear acceptor sites, cell-free nuclear binding assays with the estrogen receptor had to be developed using modifications[142] of the method used for the progesterone receptor sites.[68,69,112] Additional problems may be encountered when other steroids are used

Table 1

EVIDENCE SUPPORTING THE BIOLOGICAL RELEVANCE OF THE CELL-FREE BINDING ASSAYS

The receptor binds the steroid under cell-free conditions with the same affinity and apparent specificity as measured under whole cell conditions[2-5,35,36,91,112,121,149,166,191-193] [a]

The steroid-receptor complex formed under cell-free conditions has the same physicochemical properties (molecular weight, hydrodynamic properties, etc.) as that complex formed in intact cells[2-5,35,36,91,121,149,166,191-193]

Requirements for nuclear uptake and binding in cell-free assays (e.g., steroid bound to a receptor, receptor activation, etc.) are the same as those for the whole cell[2-5,32,35,81,91-94,112,121,149,191-193]

Conditions required for activation of the receptor, a prerequisite for nuclear uptake and binding, are the same for both cell-free and whole cell systems[2-5,35,37,81,91,112,121,149,191-193]

Under proper conditions, the cell-free binding results in a similar pattern and the level of nuclear binding as does the whole cell binding;[1,4,6-14,17-21,23,24] levels of both cell-free and whole cell nuclear bound steroid correlate with physiological responses[68,69,154,195,196]

The properties of the triplex of steroid-receptor-chromatin formed under cell-free conditions closely resembles that formed under whole cell conditions with respect to dissociation by salt, by certain divalent ions, and detergents, etc.; the dissociated radioactive steroids are still complexed to the receptor[1-5,81,91-94,112,121,149,166,191-193]

The interaction of an isolated steroid-receptor complex with isolated nuclei has been reported to alter RNA polymerase activity and transcription of selected genes in a pattern similar to that which occurs in the intact cell[41,143-148,152,153]

Conditions and/or periods in vivo in which the steroid receptor fails to translocate and bind to nuclear acceptor sites and alter transcription; this includes the annual rhythms in receptor function in the nature avian oviduct, as well as the lack of progesterone function in the immature oviduct and the oviduct of molting hens[39,117,204]

Specific nuclear acceptor sites for estrogen and progesterone receptors have been demonstrated in vivo as well as in the cell-free assays[203]

[a] References are numbered according to reference list of entire chapter.

Taken from Thrall, C. L., Webster, R. A., and Spelsberg, T. C., in *The Cell Nucleus*, Vol. 6, Busch, H., Ed., Academic Press, New York, 1978, 461. With permission.

such as androgens and glucocorticoids which can interact with the receptors of other steroids and with serum binding protein.

D. Biological Relevance of the Cell-Free Nuclear Binding of the Progesterone Receptor

Before discussing the isolation and characterization of the nuclear acceptors, a brief discussion of the biological relevance of the cell-free assays should be made. Table 1 compares the properties of cell-free nuclear binding assays, which have been reported by many laboratories for many steroid receptor-target tissue systems, to the native nuclear binding.

1. Alteration of Transcription in Cell-Free Binding Assays for Steroid Receptors

Many laboratories have reported that the cell-free nuclear binding assays alter transcription of the DNA similar to the in vivo conditions.[41,143-153] These studies do support the nativeness of the cell-free nuclear binding assays. Unfortunately, none have included proper controls in their assays. Most of these studies have not been pursued, so whether these responses in vitro occur by the same mechanism(s) as occurs in vivo remains dubious.

2. Quantitative Correlations Between In Vitro and In Vivo Nuclear Binding Sites

Several lines of evidence support the view that the cell-free nuclear binding in the progesterone receptor (P-R) avian oviduct system represents the native (endogenous) binding sites. It has been reported using whole cell and cell-free studies that multiple

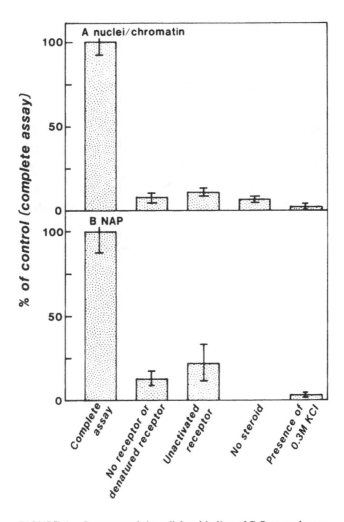

FIGURE 6.   Summary of the cell-free binding of P-R to nuclear acceptor sites. Panel A shows the level of P-R binding to oviduct nuclei or chromatin under the varying conditions listed. Panel B shows the levels of P-R binding to oviduct NAP under similar conditions. The cell-free P-R nuclear binding was performed as described elsewhere.[81,201]

classes of nuclear binding sites exist for P-R.[65,67-69,81,112,154] In these studies, the binding to the highest affinity class of sites occurred when physiological levels of the steroid in the plasma are achieved.[68,69,154] As shown in Figure 3, these high affinity sites but not the low affinity sites survive conditions of physiological ionic strengths in the cell-free binding assay.[65-69,112] The high affinity nuclear binding sites measured in the cell-free binding assays have been shown to quantitatively resemble those measured in the nuclei with in vivo studies.

3. Qualitative Correlations Between the In Vitro and In Vivo Nuclear Binding
a. Receptor Dependency of the Steroid Binding

As shown in Figure 6 (Panel A), the high affinity sites measured in the cell-free assays also resemble those measured in vivo with regard to a dependency on an intact steroid-receptor complex and to the dissociation of the steroid-receptor complex from the chromatin binding sites by 0.3 M KCl.[81,112]

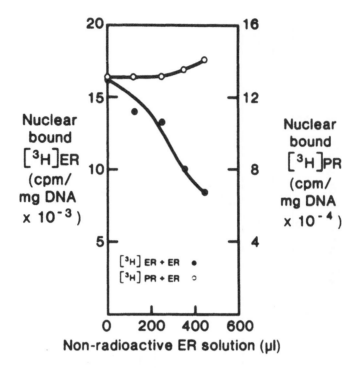

FIGURE 7.   Competition of nonradioactive ER for binding of [³H]ER and [³H]PR to isolated oviduct nuclei. A constant volume (400 μl) of [³H]PR prepared as previously described[112] or of [³H]ER prepared as previously described[142] (with an additional heparin agarose chromatography) was added to the cell-free nuclear binding assays. Increasing amounts of unlabeled ER was then added. In order to maintain constant protein concentration in all incubations, the volume of each incubation was brought to 1 ml with redissolved ammonium sulfate precipitate (35% saturation) of spleen cytosol prepared in a similar manner as ER. The ratio of [³H]receptor complex to nuclei employed in this experiment was saturating, as previously determined. Nuclear bound [³H]ER in the presence of nonradioactive ER is represented by (●); nuclear bound [³H]PR in the presence of nonradioactive ER is represented by (O). Each value is the mean of triplicate determinations.[203]

b. Receptor Specificity of the Nuclear Binding

A receptor specificity is observed for nuclear binding sites measured in the in vivo and in vitro assays. In these studies the estrogen receptor from the chick/hen oviduct was isolated, partially purified, and sufficiently stabilized for use in a cell-free binding assay.[142] Figure 7 shows that using competitive binding studies in the cell-free assay described here for P-R, the unlabeled estrogen receptor effectively competes with the radiolabeled estrogen receptor but not with the radiolabeled P-R.[203] Studies in vivo support the cell-free binding assays in that the progesterone does not compete with [³H]estrogen for nuclear binding and vice versa.[203] These studies support those of Higgins et al.[94] who reported receptor specific acceptors for estrogen and dexamethasone in the rat uterus and HTC cells. The observation that the cell-free assay demonstrates receptor-specific acceptor sites, a specificity observed in vivo, provides yet further support for the nativeness of these assays.

c. Correlations in the Patterns of Binding of Steroid Receptors Between In Vitro and In Vivo Nuclear Binding

Finally, recent studies in the author's laboratory has revealed a marked correlation between the endogenous nuclear binding and the cell-free binding. In one instance, a marked correlation between the cell-free and endogenous nuclear bindings were observed at different periods of the year. Every winter for a 4-year period a loss of nu-

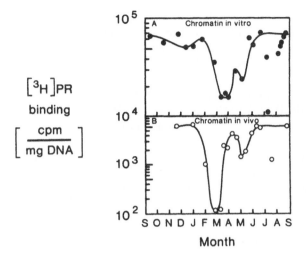

$[^3H]PR$

binding

$$\left[\frac{cpm}{mg\ DNA}\right]$$

FIGURE 8.   Seasonal variations in the capacity of [³H]PR to bind to chromatin acceptor sites in vivo and in vitro. Panel A represents the in vitro binding to whole chromatin performed within 1 week in June, using [³H]P-R isolated at various periods of the year. The receptor preparations and the nuclear binding assays were conducted as described elsewhere.[39,67,112] Saturating levels of [³H]P-R (300 μℓ/assay) for 60 μg/DNA assay were used in the binding assays. Each point represents the mean of four replicate analyses of the binding of [³H]P-R to chromatin. The receptor preparations were isolated at specific times during the year, stored at −80°C as ammonium sulfate precipitates, and resuspended on the day of the binding assay. The chromatin represented two preparations which were stored similarly and used upon demand. Panel B represents the in vivo binding of [³H]P-R to nuclear chromatin in fully developed oviducts of immature chicks conducted at various periods of the year. DES-treated chicks were injected with 200 μCi of [³H]P in 50 μℓ of ethanol-H₂O (1:1) into the wing vein. Evans Blue dye was included as a marker for the accuracy of the injection. One half hour after injection, the birds were sacrificed and the oviducts quickly excised. The nuclear chromatin was immediately isolated and quantitated for DNA and assayed for nuclear binding. The nuclear bound [³H]P was extracted with 0.3 M KCl and quantitated by charcoal analysis. The extracted radioactivity was found to be bound to a 4S sedimenting macromolecule in sucrose. The blood was collected and the radioactivity per 50 mℓ did not change. The data are plotted as CPM bound per milligram DNA for the nuclear binding vs. the date on which [³H]P was injected. (Reprinted with permission from Boyd, P.A. and Spelsberg, T. C., *Biochemistry*, 18, 3685. Copyright 1979 American Chemical Society).

clear binding capacity of P-R to whole oviduct chromatin and NAP was demonstrated.[39,117] Figure 8 shows the binding of P-R to oviduct chromatin in vitro (Panel A) and in vivo (Panel B). Clearly, a seasonal variation in the pattern of binding is observed with a decrease or loss in binding occurring in the winter. These results are further supported by a similar seasonal pattern in the effects of progesterone on RNA polymerase II activity (transcription), representing the first nuclear response to the binding of the steroid-receptor complex to nuclear acceptor sites.[39,117] These circannual rhythms were also accompanied by similar rhythms in the oviduct weights and cytosol protein. The rhythm in the nuclear binding was subsequently found to be due to functional changes in the progesterone receptor.[39,120]

Similar correlations between the in vivo and in vitro nuclear binding have also been shown in the developing oviduct. In the immature oviduct, the progesterone receptor is not capable of translocating and binding to nuclear acceptor sites and altering transcription. As oviduct development progresses, this capacity gradually appears. The cell-free binding assays similarly show that the progesterone receptor from the immature oviduct displays no nuclear binding while the receptor from the more developed stages does show nuclear binding.

In summary, the cell-free nuclear bindings mimic the endogenous nuclear binding which in turn correlate with the subsequent transcriptional response to the steroid.

When the steroid-receptor complex fails to translocate and bind to nucle ar acceptor sites in vivo, the receptor fails to bind to isolated nuclei in the cell-free assays. Reciprocally at a period when the complex does bind to nuclear acceptor sites in vivo, the receptor shows marked nuclear binding in the cell-free assay.

V. NONHISTONE CHROMATIN PROTEINS MASK MANY NUCLEAR ACCEPTOR SITES FOR THE AVIAN OVIDUCT PROGESTERONE RECEPTOR

Initial studies on the effects of protein dissociation from the chromatin on the acceptor activity met with unexpected results. In the procedure of selectively removing proteins beginning with whole chromatin and ending with pure DNA, the extent of steroid receptor binding first markedly increases followed by an even greater decrease to a level of almost no binding. This rise in binding was termed an "unmasking" of acceptor sites while the subsequent loss in binding has been termed the dissociation of nuclear acceptor components from the DNA. This is described below.

A. Role of a Specific Chromatin Nonhistone Protein(s) in Masking Nuclear Acceptor Sites

In the early studies, crude fractions of the chromosomal proteins were removed from chromatin followed by an analysis of the acceptor activity on the residual protein-DNA complexes.[66,81] These studies were performed on free chromatin or chromatin attached to cellulose. Table 2 presents an outline of these fractions, how they were dissociated from chromatin, and the proportion of total chromatin protein they represent. The terminology of the residual deoxyribonucleoprotein after removal of each of the protein fractions from the chromatin is also given. Figure 9 shows the acceptor activity in chromatin (attached to cellulose) from which were removed the histones (CP-1), then the bulk of the nonhistone proteins (CP-2), and finally the remainder of the nonhistone protein tightly bound to the DNA (CP-3) to yield pure DNA.[66,81] Figure 9, Panel A show that the removal of the CP-1 fraction (histones) from oviduct chromatin causes little change in the binding of P-R to the residual nucleoprotein (dehistonized chromatin). When fraction CP-2 (representing the bulk of the nonhistone protein) is removed, however, the binding by the P-R is markedly increased.[1,66,68,69,81] Finally, the removal of the CP-3 fraction yields pure DNA which displays a low level, nonsaturable binding. Table 3 shows the extent of masking in oviduct chromatin with regard to the number of molecules of the P-R bound to nuclear acceptor sites per cell.

Following lengthy studies of this phenomenon (discussed further in the following sections), the increase in binding was termed an "unmasking" of acceptor sites with the CP-2 fraction assumed to contain the masking factor(s). As shown in Table 2, the residual DNA-CP-3 complex containing the high capacity binding has been termed "nucleoacidic protein" or NAP. The decrease in binding of the progesterone receptor with the removal of the CP-3 fraction suggested that this fraction contains the acceptor component(s). This CP-3 fraction is discussed in detail later.

Interestingly, the nontarget tissue chromatins such as spleen and erythrocyte show little or no binding by the P-R (Figure 9, Panel B).[1,65,66,68,69,81,111] When the CP-1 and CP-2 fractions are removed from these chromatins, the resulting NAPs display marked binding, equivalent to the NAP obtained from the target tissue chromatin. Thus, the acceptor sites in nontarget tissue chromatin appear to be totally masked, i.e., are not expressed. When the NAPs from these chromatins are completely deproteinized, the resulting pure DNAs show the same low level of binding as the DNA from oviduct. Table 3 shows the extent of masking to these nontarget tissue chromatins based on the number of steroid-receptor complexes bound to nuclear acceptor sites per cell. Thus,

Table 2
CRUDE FRACTIONATION OF PROTEINS FROM THE AVIAN OVIDUCT CHROMATIN

Residual complex of deoxyribonucleoprotein (after removal of fraction)	Solvent used to extract protein	Protein fractions removed	Abbreviated name of fractions	Quantity of fraction in chromatin (mg/mg DNA)
Chromatin	None	None	None	—
Dehistonized chromatin	3.0 M NaCl pH 6.0	All histones 10% nonhistone	CP-1	1.2
Nucleoacidic protein (NAP)	4.0 M GuHCl pH 6.0	75% of nonhistone protein	CP-2 (masking proteins)	1.4
DNA	7.0 M GuHCl pH 6.0	15% of nonhistone protein	CP-3 (acceptor proteins)	0.2

Data modified from Webster, R. A., Pikler, G. M., and Spelsberg, T. C., *Biochem. J.*, 156, 409, 1976.

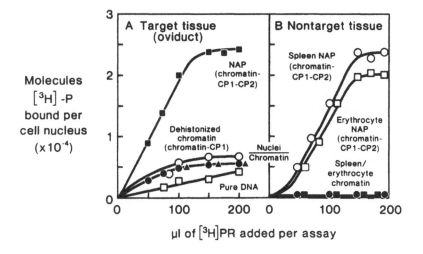

Molecules
[^{3}H]-P
bound per
cell nucleus
($\times 10^{-4}$)

μl of [^{3}H]PR added per assay

FIGURE 9. Masking of nuclear binding sites for the progesterone receptor in the avian oviduct. The in vitro binding of [^{3}H]P-R to the soluble DNA or DNA-protein complexes was performed using the cellulose method. The high affinity binding in vitro was assayed in 0.15 M KCl using 0 to 200 μl of labeled receptor and 50 μg DNA (as chromatin or protein-DNA complexes). The removal of total histones (CP-1), the first group of nonhistone proteins (CP-2), or the remaining nonhistone protein (CP-3) has been described elsewhere.[66,68,69] The efficiency of removing these fractions from chromatin DNA and the integrity of the nuclear components (histones) were assayed by polyacrylamide gel electrophoresis and by estimation of the protein-DNA ratio. In Panel A the symbols represent binding to the following for each tissue: (▲) nuclei; (●) whole chromatin; (○) chromatin deficient of histone (CP-1); (■) chromatin deficient of CP-1 and CP-2, termed nucleoacidic protein or NAP; and (□) chromatin deficient of total protein (pure DNA). From the [^{3}H]P-R bound per milligram DNA was calculated the molecules of nuclear bound P-R per cell. In Panel B (●) represents spleen chromatin, (■) mature erythrocyte chromatin, (○) spleen NAP, and (□) erythrocyte NAP which represent chromatins devoid of CP-1 and CP-2. (Reprinted by permission from Spelsberg, T. C., Webster, R. A., and Pikler, G. M., *Nature (London)*, 262, 65. Copyright © 1976 Macmillan Journals Limited.)

the chromosomal material of many tissues of the bird appear to contain the same number of acceptor sites for the progesterone receptor but the majority in the oviduct and all in nontarget tissues are "masked". Similar findings have subsequently been reported in other steroid target tissue systems.[80,82,83,205,206] It should be mentioned that Chytil and Spelsberg[155,156] successfully prepared rabbit antisera to the chick oviduct

Table 3
EXTENT OF MASKING OF HIGH AFFINITY SITES[a]

Source	Level of binding at saturation (molecules/cell)	Percentage of total sites masked
Oviduct nuclei in vivo	10,600	58
Oviduct nuclei in vitro	5,951	76
Oviduct chromatin in vitro	7,441	71
Oviduct chromatin minus histone (CP-1)	9,278	58
Oviduct chromatin minus histone (CP-1) and CP-2	25,290	0
DNA	2,055	—
Spleen nuclei	60	100
Spleen chromatin	48	100
Spleen chromatin minus histone (CP-1) and CP-2	24,106	0
Erythrocyte nuclei	357	100
Erythrocyte chromatin	416	100
Erythrocyte chromatin minus histone (CP-1) and CP-2	15,018	0

[a]   The number of binding sites measured in each of the chromatin minus histone (CP-1) and CP-2 preparations was assumed to represent the total sites in the respective chromatin, and thus each of these preparations was assigned 0% masking. The values at saturation for each preparation were taken at 100 μg of injected hormone for the in vivo binding and 200 μℓ labeled receptor preparation for the in vitro binding. The preparations of the nuclear components as well as the hormone binding assays were performed as described elsewhere.[81]

CP-3 protein fraction. Using a complement fixation assay, these investigators demonstrated that the majority of the antigenic sites of the CP-3 fraction were "masked" in whole oviduct chromatin. Removal of the CP-2 fraction unmasked 80% of the antigenic sites. These results using the complement fixation assays with antisera to CP-3 closely mimic those obtained with the cell-free nuclear binding assays of P-R with regard to the "masking" of the CP-3 proteins.

The CP-2 proteins have been reannealed to the NAP (the DNA-CP-3 complex) using a regressing concave gradient from 6 to 0 *M* GuHCl in specially designed chambers.[202] Interestingly, a remasking of a similar quantity of sites as measured in the original intact chromatin is achieved. The CP-2 fraction has been subfractionated into three fractions, CP-2a, CP-2b, and CP-2c. As shown in Figure 10, when a variety of different protein fractions are reannealed to NAP, only one fraction, the CP-2b, representing only 10% of the CP-2 fraction, displays the masking activity by markedly lowering the extent of binding of the P-R. The CP-2a fraction, ovalbumin, and histones display no masking activity. The increase in binding observed with the ovalbumin was observed occasionally and found to be caused by a protection of the acceptor sites from proteolytic activity occurring toward the end of the reconstitution. Since proteolytic but not nucleolytic action on chromatin causes an unmasking of sites (see Figure 5), protein(s) are believed to represent the masking factors.[202] Since the CP-2b fraction dissociates from chromatin long after the histones and since the histones do not mask

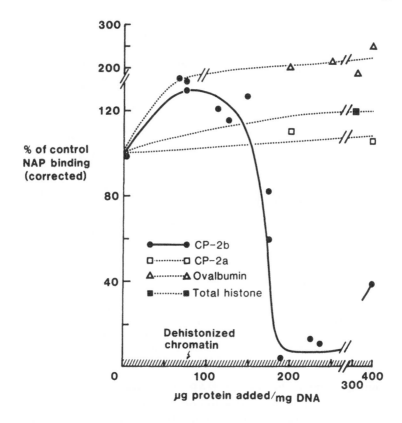

FIGURE 10.   Reconstitution of chromatin components which mask nuclear acceptor sites for progesterone. The CP-2 fraction of the chromatin proteins was subfractionated into several groups: CP-2a, CP-2b and CP-2c, representing 10, 6, and 33% of the total chromatin protein, respectively. Two of these fractions, as well as standard ovalbumin and total histone, were reannealed to the NAP at increasing ratios of protein to DNA using methods described previously.[6,67-69] After reconstitution, the DNA bound protein was separated from the unbound protein and the former assayed for P-R binding using the streptomycin assay.[81] The values are corrected for DNA binding. (●) CP-2b, (□) CP-2a, (△) ovalbumin, and (■) histone. The hatched area on the abscissa represents values for P-R binding to the dehistonized chromatin.[202]

the acceptor sites in the reconstitution assays, it is speculated that the masking activity is a specific class of nonhistone proteins. Preliminary analysis by molecular sieve chromatography using CL-Sepharose® 6B in 6 *M* GuHCl containing mercaptoethanol (pH 6.0) suggests a very heterogeneous population of masking "proteins" with monomer molecular weights ranging from 60,000 to 130,000 and possibly larger.[202]

B. Properties of the Masked Acceptor Sites: Comparison with the Unmasked Acceptor Sites in Whole Chromatin

In order to examine the significance of these findings, the biological relevance of these masked sites was examined. Comparisons between the binding of the progesterone receptor to the oviduct NAP (in which all acceptor sites are exposed) and the binding to whole chromatin (in which only a fraction of the sites are exposed) were performed.

Figure 6 (Panel B) shows that the binding of the progesterone receptor to acceptor sites on the oviduct NAP (containing all acceptor sites including many which normally are masked in whole chromatin) requires an intact, activated receptor as found with the unmasked sites in whole chromatin. Further, the receptor is dissociated from the NAP with 0.3 *M* KCl as in the case for the binding to unmasked sites. Figure 9 shows

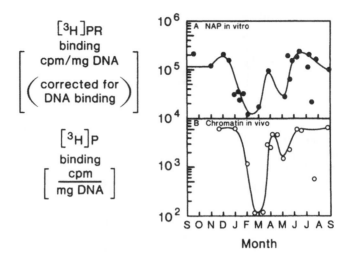

$$\left[^3H\right]PR$$
binding
cpm/mg DNA
$$\left(\begin{array}{c}\text{corrected for}\\\text{DNA binding}\end{array}\right)$$

$$\left[^3H\right]P$$
binding
$$\left[\dfrac{cpm}{mg\ DNA}\right]$$

FIGURE 11.   Seasonal changes in P-R binding to nuclear acceptor sites in the chick oviduct. The binding of P-R to isolated NAP in vitro and chromatin in vivo using receptors isolated at various times of the year has been described in the legend of Figure 8 and elsewhere.[39,120] Panel A shows the seasonal binding by P-R to NAP (representing the total acceptor sites) in vitro. Panel B shows the seasonal binding to whole oviduct chromatin in vivo. (Reprinted with permission from Boyd, P. A. and Spelsberg, T. C., *Biochemistry*, 18, 3685. Copyright 1979 American Chemical Society.)

that the binding to these sites on the NAP are saturable. Previous studies showed similar affinities of binding of the P-R to both the masked and unmasked sites with a $K_d \sim 10^{-8}$ to 10^{-9} M.[81] Thus, a marked similarity between the acceptor sites expressed in intact chromatin (representing unmasked sites) and the sites on NAP (representing both masked and unmasked sites) is found.

The receptor preparations showing seasonal rhythms in nuclear binding as described earlier and in Figure 8 were then used to analyze these unmasked sites. Figure 11 shows that cell-free binding of the progesterone receptor to the NAP (with all binding sites expressed) reflects the same seasonal rhythm as does the in vivo binding to whole chromatin (with 80% of the binding sites masked). Similar studies were performed using the estrogen-induced oviduct development system. As shown in Figure 12, in the early stages of the oviduct development, the P-R is incapable of translocating and binding to nuclear acceptor sites in vivo and altering transcription. As development progresses, the receptor acquires the capacity for nuclear binding and inducing changes in transcription.[204] Interestingly, the P-R isolated from the undeveloped oviduct also shows little or no cell-free binding to NAP. As development progresses, these isolated receptors gradually gain the capacity for binding to NAP in the cell-free assays. These changes in the cell-free binding are not observed when pure DNA or DNA-CP-1 fractions are used in place of NAP. Since the NAP used in these studies was the same preparation from mature oviducts, the pattern of binding was not due to the nuclear acceptor sites. Thus, the cell-free binding to NAP mimics the binding to chromatin in vivo.

In conclusion, it can be seen that the normally masked sites in chromatin which are unmasked by the removal of the CP-2 fraction (to yield NAP) appear to have many properties in common with the unmasked sites on the whole chromatin. The similarities include dependency on an intact, activated receptor, the affinity of binding, the concentration of salt required to dissociate the steroid-receptor complex, and the patterns of binding observed during oviduct development and during the year in the mature oviduct. Absolute proof of identical sites will require purification of the acceptor pro-

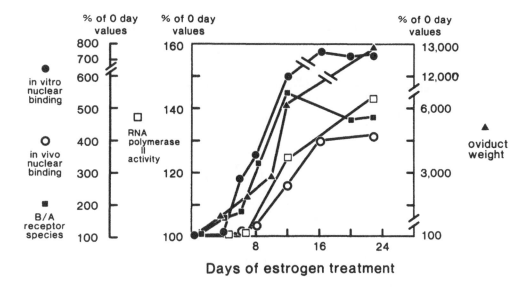

FIGURE 12. Composite of functions of the progesterone receptor during oviduct development. The undeveloped oviducts of immature chicks were induced to full development with daily injections of estrogens. Changes in the (▲) oviduct weight, (●) cell-free NAP binding by the isolated P-R, (■) ratio of the quantity of the B species and the A species of the P-R, (O) nuclear binding in vivo, and (□) RNA polymerase II responses to injected progesterone were analyzed. The methods are described elsewhere.[39] It can be seen that a nonfunctioning P-R (i.e., incapable of nuclear binding and alteration of transcription) occurs in the immature oviduct, explaining the nonresponsiveness of the organ to progesterone.[204]

teins, preparation of antisera, and comparison of antigenic properties between the masked and unmasked sites.

C. Regulation of Acceptor Site Expression by Masking Proteins

The question arises as to the biological role of the masking phenomenon. One possible function may be that the masking regulates which genes will respond to the P-R. The regulation of which genes will respond to a steroid would explain a heretofore unexplained but important dilemma in endocrinology: "why different target tissues of the same organism with presumably the same type of receptor display markedly different responses with regard to gene expression to the same steroid." In short, different genes seem to be regulated by the same steroid-receptor complex in different tissues. Recent evidence indicating that the steroid receptors are antigenically similar in different tissues of different animals minimizes the tissue specificity of the receptors themselves.[157-159]

In short, those genes which are masked in the oviduct chromatin may represent either genes which are currently expressed (unmasked) in other progesterone target tissues or genes which were expressed at an earlier stage of development of the organism. In nontarget tissues which do not possess steroid receptors and thus do not require genes that respond to steroids, all such steroid-regulated genes would be masked. This is supported in Figure 9B wherein all acceptor sites appear to be masked in the chromatin of a nontarget tissue. This concept is further supported by studies on the developing oviduct. During the development of a target organ such as the chick oviduct, quantitative and qualitative changes in the masking activity might be expected as different cell types appear or change in proportions. During oviduct development, different genes are found to respond to the same steroid such as estrogen.[160] Figure 13 shows that quantitative changes in the extent of masking of a constant amount of acceptor sites occurs during oviduct development.[202] The number of available acceptor sites

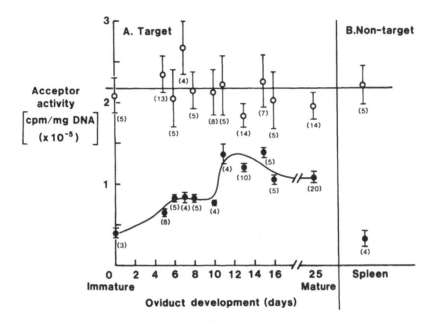

FIGURE 13.   Differential masking of progesterone acceptor sites during oviduct cytodifferentiation. Panel A shows the acceptor levels for P-R (at saturation) in (●) whole
chromatin or (○) NAP during oviduct development using a cell-free binding of P-R as
described elsewhere.[39,81] Panel B shows the same values for spleen, a nontarget tissue
for progesterone removed at the 25th day of estrogen treatment. The mean and standard deviation of multiple replicate analyses of the binding (the number shown in paren-
theses) are presented. These experiments are described in greater detail in the text.[202]

(chromatin binding) but not the total number of sites (NAP binding) changes throughout the estrogen-induced development. Thus, the extent of masking changes considerably.

Indications that the masking during organ development may be qualitatively changing was suggested by antigenic analysis of the CP-3 fraction during rat liver development. Using the complement fixation assay with antisera prepared against the CP-3
fraction of the adult rat liver chromatin, as discussed earlier with the chick oviduct,
[155,156] an indication of possible changes in masking during the development of the rat
liver was observed.[161] One explanation of these results is that the masking of different
fractions of the CP-3 proteins is occurring on the chromatin during organ development.

Further, several other laboratories have reported similar masking of nuclear acceptor sites of other steroid receptor systems. These are androgens in the rat prostate,[80]
estrogens and progesterone in the sheep brain,[82] glucocorticoids in the rat liver,[83] estrogens in the rat and bovine uterus,[205] and progesterone in the guinea pig uterus.[206] The
basic results of these studies in general are very similar to those described in this chapter for the avian oviduct progesterone receptor.

Therefore, the masking phenomenon may well represent the means by which steroid-
regulated genes are differentially expressed in different target tissues. At present this
model of masking remains to be verified. Evidence that masking exists in other steroid
target tissue systems lends support to its existence. The evidence of total masking of
the steroid receptor binding sites in nontarget tissues is interesting but the rationale
for it is obscure since these tissues do not contain receptors anyway.

D. A Hypothetical Model for the Masking Process

Figure 14 shows a model of chromatin with its nucleosomes containing the bulk of

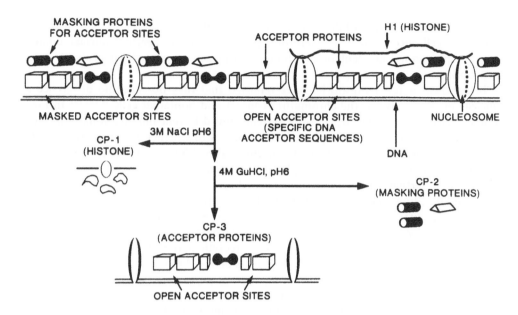

FIGURE 14. Model of the oviduct chromatin with regard to unmasking of acceptor sites in avian oviduct chromatin. This model depicts the DNA wrapping in and around the nucleosomes (Nu bodies) which contain all of the histones, termed the CP-1 fraction, except the H1 histone species. The multitude of different nonhistone species are represented in large groups as symbols, i.e., one symbol represents many species. The cylinders represent masking proteins which are a part of the CP-2 fraction which in turn masks the acceptor proteins, represented by the squares. The squares which represent the acceptor proteins are a part of the CP-3 fraction. These acceptor proteins in turn sit on specific DNA sequences. Positions on the intact chromatin where the cylinders do not cover the squares represent the "open acceptor" sites for P-R; those with cylinders covering them represent masked sites. The model also depicts the removal of the CP-1 fraction with 3 M NaCl and the CP-2 fraction with 4 M GuHCl. The latter causes an unmasking of all acceptor sites, as depicted in the figure.[201]

the histones, a suggested position for the H-1 histone, and expressed (unmasked) acceptor sites (the square blocks) as well as the masked acceptor sites (square blocks covered overlaid with cylinders). The acceptor sites (denoted as blocks) are bound to specific DNA sequences (discussed later). The removal of the CP-1 fraction (or histones) by 3 M NaCl does not remove the masking proteins. The removal of the CP-2 fraction, represented by the cylinders, with 4 M GuHCl to yield the NAP does remove the masking proteins (and other proteins) and causes a marked enhancement in binding of the P-R by exposing many additional acceptor sites. During cytodifferentiation, the masking components (the cylinders) would be shifted from one acceptor site to another, thereby preventing some while allowing others to bind to incoming steroid receptors and respond to the steroid stimulus.

E. The Role of Masking in the Controversy Over Tissue-Specific Acceptor Sites

There is a controversy as to whether or not there is tissue specificity with respect to the nuclear acceptor sites for steroid receptors on chromatin. Since nontarget tissues do not have cytoplasmic receptors for steroids, whether or not their chromatin contains acceptor sites for the steroid is, biologically speaking, of no consequence. The question arose from studies of cell-free binding assays wherein heterogeneous assays could be generated using chromatin from nontarget tissues and steroid receptors from target tissues. Numerous laboratories have reported that a variety of target cell nuclei for a variety of steroids contain markedly more acceptor sites than nontarget cell nuclei using cell-free binding assays.[1,36,37,64-71,76,81,92-94,112,114,162-172] Many of these reports did not state that nontarget cell nuclei lacked any acceptor sites, but that they contained fewer numbers of them, while some reported a total lack of acceptor sites in certain nontarget cell chromatin.[81,111] Therefore, the term "tissue-specific" nuclear binding may be misleading; an "enhanced" binding in target tissue nuclei is probably more appropriate.

In contrast, a few laboratories have reported that they observed no difference in nuclear binding between the chromatin of target and nontarget cells.[94,140,141] In these studies, different steroid-target tissue systems and different binding conditions were used. These factors probably explain the discrepancy of whether or not a tissue-specific nuclear binding exists on whole nuclei or chromatin. First of all, the role of the free steroid and the ionic conditions can play a significant role in the extent of the cell-free binding to a particular chromatin, as discussed earlier. The masking of a multitude of acceptor sites, however, may also explain the discrepancy in the reports on tissue-specific acceptor sites. As described earlier and in Figure 5, our studies have revealed that even very mild protease treatment of oviduct chromatin results in the unmasking of acceptor sites.[202] The same unmasking by proteolytic action occurs in the chromatin of nontarget tissues. Since proteolytic activity is detected in both steroid receptor preparations and in the nuclei or chromatin, it is very possible that some studies were performed under conditions allowing mild proteolysis to occur. As discussed in Section IV.B.1 and IV.C.1, when [14C]-ovalbumin is incubated in the cell-free assay and subsequently analyzed on SDS-polyacrylamide gel electrophoresis, proteolytic breakdown of the ovalbumin is observed under many conditions.[201] Indeed, conditions that enhance proteolysis such as use of crude cytosol, temperatures above 4°C and long periods of incubation were used in the studies reporting a lack of tissue-specific acceptor sites on chromatin.[94,140,141] Thus, the chromatin from nontarget tissues, which normally may display no steroid receptor binding (i.e., no exposed acceptor sites), when subjected to mild proteolysis during cell-free binding or even during isolation, do generate significant binding of the steroid receptor. These results have been misinterpreted as a lack of tissue-specific acceptor sites, but actually the specificity exists.

F. Role of Masking in the Controversy Over Saturable Binding of Acceptor Sites in Cell-Free Binding Assays

Unsaturable binding to nuclear acceptor sites has been reported for a couple of steroid-target tissue systems using the cell-free binding assays. While some of these studies can be explained by low ionic conditions,[111] other studies using estradiol[29] and glucocorticoids[139] included adequate ion concentrations to identify the saturable, high affinity binding sites. As discussed in Section IV.C.4, recent studies in this laboratory revealed that the high adsorptivity of certain steroids such as estrogen causes significant nonspecific, nonsaturable binding to nuclei in cell-free assays.[142] The evidence for proteins masking acceptor sites and the action of proteases on this masking gives an additional or alternate explanation for the apparent nonsaturable binding in the cell-free assays.

As shown in Figure 5, very mild proteolytic action on oviduct chromatin causes a rise in the binding of the P-R. Based on the evidence discussed in Section V.E, the protease is degrading the masking proteins, causing an increase in steroid receptor binding in the previous section. As discussed above, analysis for proteolytic activity in cell-free nuclear binding assays reveals minimal activity in assays incubated for 90 min at 4°C using partially purified receptor preparations. In contrast, incubations for periods longer than 90 min, or under higher temperatures, or in the presence of a crude receptor preparation results in significant proteolytic activity and actually causes an increase in P-R binding.[202] Therefore, the conditions in a cell-free binding assay can readily cause an increase (i.e., an unmasking) of acceptor sites to create the appearance of a nonsaturable binding to nuclear acceptor sites.

Table 4
IMPORTANT CONSIDERATIONS IN THE RECONSTITUTION OF THE ACCEPTOR "ACTIVITY" OF THE AVIAN OVIDUCT PROGESTERONE RECEPTOR

A special chamber is used; the chamber contains inlet and outlet valves; the dialysis bags with the
reconstitution mixture are placed in the chamber which sits on a rocking platform; the 6 M GuHCl in the
chamber is gradually replaced by buffer using a peristaltic pump while the chamber is being rocked; air
bubbles perform the mixing both inside and outside the dialysis bags
The following conditions were found to markedly affect the success of the reconstitution
    The concentration of the DNA ($\sim$0.5 mg/mℓ is best)
    The ratio of the protein and DNA (discussed in Section VI.B)
    The pH of the solvents ($\sim$pH 6.0 is best)
    The period of the reverse gradient ($\sim$10 hr)
    The removal of unbound protein and DNA from the reconstituted nucleoacidic protein
Other conditions such as the presence of small concentrations of reducing agent and EDTA were found to
play only a moderate effect on a successful reconstitution

From Spelsberg, T. C. and Toyoda, H., in preparation.

VI. CHEMICAL CHARACTERIZATION OF THE NUCLEAR ACCEPTOR SITES FOR THE AVIAN OVIDUCT PROGESTERONE RECEPTOR

A. Role of Specific Chromatin Nonhistone Proteins and DNA in the Nuclear Acceptor Sites
1. Method of Reconstituting the Acceptor Activity on DNA

The procedure for reannealing the isolated acceptor proteins to pure DNA was developed in part empirically and in part from the literature published over the past 20 years. Many methods used have been published which allow the refolding of denatured enzymes and proteins to their biologically active forms,[173-181] as well as to recombine proteins and DNA by methods for the reconstruction of native-like deoxyribonucleoproteins.[1,6,64,67,70,116,132,181-189] The methods and conditions used to reconstruct nucleoproteins are similar to those used to renature proteins.

Since many of the isolated chromatin proteins are hydrophobic, the activity was originally reannealed to DNA to render a more soluble complex with which to assay acceptor activity (described in detail later). Since NaCl-urea and GuHCl were used to dissociate the acceptor activity from the DNA, a method was developed following the classic reverse gradient techniques.[6,64,68,132,182-190] The method was modified for achieving optimal amounts of active acceptor sites[6,67-69,116,120] and several conditions were found to be critical. These are outlined in Table 4. Details of the method will be described elsewhere.[207] This method allows the monitoring of the acceptor activity for P-R after it is dissociated from the chromatin for characterization and purification. During the reannealing of the CP-3 fraction to pure DNA, the acceptor activity is reconstituted to the DNA when the GuHCl concentration decreases from 6 to 2 M GuHCl. At even lower concentrations of GuHCl, the activity decreases to some extent. This may be due to a blockade of some acceptor sites by "other proteins" and/or the action of proteases when GuHCl concentrations reaches 0.5 M or less. In any case, the acceptor activity can be reconstituted on the DNA and the method can be used during further purification steps to identify the acceptor activity.

2. Identification of the Acceptor Activity With Specific Protein Fraction

Figure 15 shows the acceptor activity remaining on the protein-DNA complexes after the oviduct chromatin was treated with various concentrations of GuHCl. In Panel A

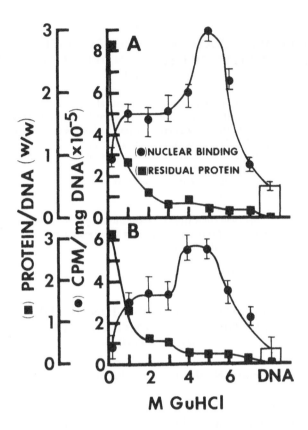

FIGURE 15.   Binding of [³H]P-R to nuclear acceptor sites on residual deoxyribonucleoprotein after extractions with sequential increases in GuHCl concentration of chromatin. Panel A represents binding of [³H]P-R to hen oviduct chromatin-cellulose treated with GuHCl; Panel B represents P-R binding to unattached hen oviduct chromatin resuspended in GuHCl and centrifuged. In these experiments, portions of chromatin-cellulose or unattached chromatin were washed twice in 20 volumes of solutions containing various concentrations of GuHCl buffered at pH 6.0. The cellulose resins were then washed in dilute Tris-EDTA buffer several times, frozen, and lyophilized. The free chromatin samples in various concentrations of GuHCl were centrifuged at 10⁵ g for 36 hr. The pellets of residual deoxyribonucleoproteins were resuspended in dilute Tris buffer at 1 mg DNA/mℓ and dialyzed vs. the buffer. The residual material was tested for (■) protein and (●) acceptor activity using saturating levels of the P-R. The binding assays were performed essentially as described by Webster et al.[81] The average and range of three replicate analyses for each assay of the hormone binding are shown. (Panel A is reproduced from Spelsberg, T. C., Webster, R. A., and Pikler, G. M., *Chromosomal Proteins and their Role in Gene Expression*, Stein, G. and Kleinsmith, L., Eds., Academic Press, New York, 1975, 153; Panel B is reproduced from Thrall, C. L., Webster, R. A., and Spelsberg, T. C., *The Cell Nucleus*, Vol. 6, Busch, H., Ed., Academic Press, New York, 1978, 461. With permission.)

chromatin was attached to cellulose while it was unattached in Panel B. After each extraction of the chromatin-cellulose resin in Panel A, the resin is washed with dilute buffers to remove traces of guanidine. The free chromatin (Panel B) is first sedimented by ultracentrifugation to separate it from the dissociated protein. The residual nucleoprotein is resuspended in dilute buffer and dialyzed against the same buffer to remove traces of GuHCl. In each method, there is an unmasking of acceptor sites on the residual deoxyribonucleoprotein with extractions of 2 to 4 *M* GuHCl. The level of unmasking achieved with the 4 *M* GuHCl (pH 6.0) is identical with that achieved using the 2 *M* NaCl 5 *M* urea (pH 6.0) shown in Figure 9. As shown in Figure 15, when the chromatin is extracted with 7 *M* GuHCl (or treated with phenol-chloroform or pronase as seen in Figure 3B), the acceptor activity markedly decreases to levels found with pure DNA. The 7 *M* GuHCl treatment thus dissociates the CP-3 fraction which appears to contain the "acceptor activity" from the DNA.

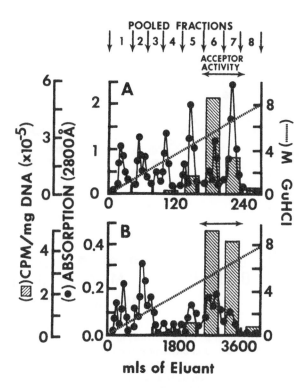

FIGURE 16. Elution of proteins and acceptor activity from chromatin-cellulose and chromatin-hydroxyl-apatite resins. In Panel A hen oviduct chromatin-cellulose resin was prepared as described elsewhere.[65,81] Briefly, 20 g of this resin containing approximately 60 mg DNA as chromatin were resuspended in 100 ml cold phosphate buffer (pH 6.0) and allowed to hydrate for 2 to 6 hr with gentle stirring at 4°C. The resin was collected in a column and a gradient of 0 to 8 M GuHCl in phosphate buffer (pH 6.0) passed through the column wthin a 4-hr period. In Panel B resins of chromatin-hydroxylapatite were prepared in the presence of 0.05 M KPO₄, pH 6.0. The resin containing 100 mg chromatin DNA and 100 g hydroxylapatite was placed on the column and 4 ml fractions were collected under a 0 to 7 M GuHCl gradient. Tubes were monitored by absorption at 280 nm. Fractions were also monitored for conductivity as well as refractive index, and the gradient level of GuHCl plotted. Fractions from both resins were pooled according to their elution with each unit of concentration of GuHCl (1, 2, 3 M etc.). Pooled samples were then dialyzed thoroughly against water and lyophilized. The lyophilized materials were resuspended in a small volume of water, homogenized in a Teflon® pestle glass homogenizer, assayed for protein, and reannealed to pure hen DNA using a reverse gradient of 6 M GuHCl to 0 M GuHCl as described in the text. The reconstituted nucleoproteins were analyzd for acceptor activity by the streptomycin method,[81] substracting the values obtained with pure DNA. The acceptor activity is presented as bar graphs. Total recoverable protein after dialysis and lyophilization was estimated to be 50% of the total protein placed on the column as chromatin-cellulose or chromatin-hydroxylapatite. (From Spelsberg, T. C., Thrall, C. L., Webster, R. A., and Pikler, G. M., *J. Toxicol. Environ. Health*, 3, 309, 1977. Copyright Hemisphere Publishing Corporation. With permission.)

The proteins dissociated from the chromatin DNA by the chromatin-cellulose and similarly by the chromatin-hydroxylapatite methods were then examined for acceptor activity. As depicted in Figure 16, the fractions eluting from these resins were pooled according to unit molarities of GuHCl in the eluants. These pooled samples were reconstituted to DNA and assayed for acceptor activity, i.e., progesterone receptor binding. Figure 16 shows that in both methods of fractionating the chromosomal proteins, the fractions dissociating from DNA between 4 and 6 M GuHCl contained the acceptor activity. Thus, as the acceptor activity disappears from the chromatin (Figure 15), it appears in the eluant (Figure 16).

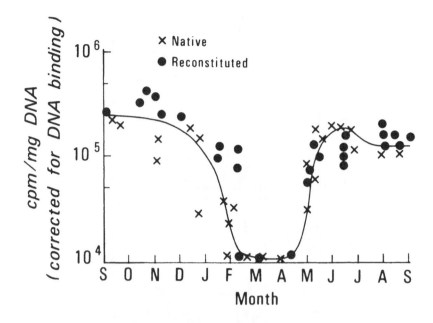

FIGURE 17.   Comparison of acceptor activities in native (undissociated) and reconstituted NAP using P-R isolated at different periods of the year. The binding of P-R isolated at various periods of the year to (x) native (undissociated) NAP and (●) reconstituted NAP were performed essentially as described in the legends of Figures 8 and 9. (Reproduced in part from Spelsberg, T. C. and Halberg, F., *Endocrinology*, 107, 1234, © 1980, The Endocrine Society. With permission.)

3. Biological Relevance of the Reconstituted Acceptor Sites

The binding of the P-R to the reconstituted NAP displays the same requirements (see Figure 6) as was determined for the native (undissociated) NAP and chromatin. Further, as shown in Figure 17, the receptor preparations isolated at various periods of the year display a seasonal rhythm in binding to a common same preparation of reconstituted NAP as is observed for the native NAP and whole chromatin (see Figures 8 and 11). These results indicate that the reconstituted NAP is very similar to the native NAP which in turn displays the same seasonal pattern of nuclear binding as found in vivo (see Figure 11). Section VI.B describing the quantitative titration of the acceptor activity on the DNA gives further support to the reconstitution of native acceptor sites.

4. Characterization of the Acceptor Protein
a. The Acceptor Factor as a Protein

Two early studies were performed to determine whether or not the acceptor activity was due to protein(s). Figure 3 (Panel B) shows that when the NAP is treated with pronase, most of the acceptor activity is lost. Figure 18 shows that, when isolated, CP-3 is subjected to increasing degrees of proteolysis (as determined by degradation of the [^{14}C]ovalbumin) and then reannealed to DNA, a loss in the acceptor function occurs. No such loss in acceptor function is observed when the CP-3 protein is subjected to ribonuclease action.[201]

A more physical chemical approach was used to show the proteinaceous property of the acceptor activity. Figure 19 shows the patterns of isopycnic centrifugation of the DNA-free acceptor "activity". Standards for protein, RNA, and DNA were also applied to similar gradients. The gradients consisted of 6 *M* GuHCl with increasing concentrations of CsCl$_2$ and run at 5°C for 72 hr. This method was developed for application to hydrophobic proteins.[201] Fractions from the gradient were collected,

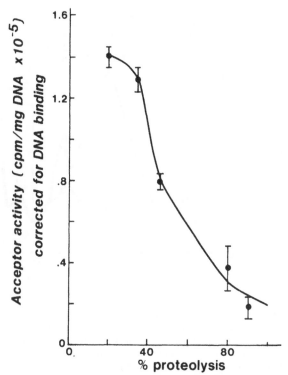

FIGURE 18. Effects of proteolysis on the CP-3 fraction on the P-R acceptor activity. The CP-3 fraction in dilute buffer was subjected to endogenous proteolytic action by incubations at 4°C over a 4-day period. [^{14}C]ovalbumin was included in the incubation solutions to serve as an indicator of proteolysis. Part of the solutions were applied to SDS-polyacrylamide gel electrophoresis, the gels sliced, and counted. The percent of the total gel radioactivity that is measured in the ovalbumin migration zone is used to calculate the percent proteolysis. The other portion of the incubation solution is reannealed to pure DNA and analyzed for P-R binding activity (i.e., acceptor activity). The P-R binding is plotted against the percent proteolysis. The mean and standard deviation of four replicate binding analyses is shown.[201]

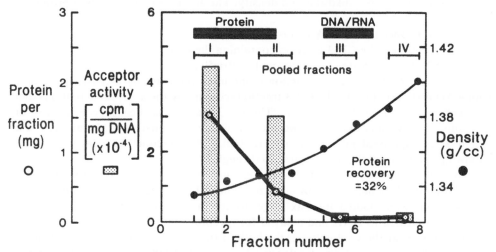

FIGURE 19. Isopycnic centrifugation of acceptor proteins in CsCl$_2$ gradients containing 6 M GuHCl. The CP-3 protein fraction was placed in 6 M GuHCl, buffered at pH 6.0, and the solution layered over a gradient of 10 to 50% CsCl$_2$ in 6 M GuHCl in SW 50.1 rotor tubes. The tubes were centrifuged for 72 hr at 100,000 × g_{ave}. Some tubes contained standard DNA, RNA, or protein. Fractions of the gradients were collected and the (●) density determined. The fractions were then pooled as depicted in the figure, and the (O) protein content quantitated. These fractions were reannealed to pure DNA and the reconstituted NAP assessed for P-R binding (acceptor activity) as illustrated by the bars. Values for P-R binding to pure DNA were subtracted from the respective NAP binding values.[201]

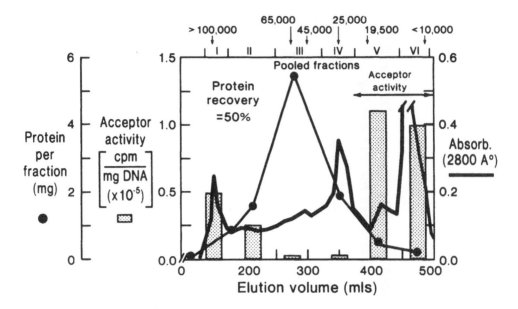

FIGURE 20.   Molecular sieve chromatography of the CP-3 proteins and acceptor activity in CL-Sepharose® 6-B. The CP-3 protein fraction was resuspended in a 6 *M* GuHCl solution at 2 mg protein per m*l*. This protein solution was clarified by centrigutaion at 20,000 × g for 10 min and then applied to a 2.6 × 94 cm column of the resin. The eluted fractions were pooled according to the absorbing peaks (—) as shown in the figure. The pooled fractions were quantitated for (●) protein and then reannealed to DNA and assayed for "acceptor" activity (as illustrated by the bars) using the streptomycin assay.[81] Values for P-R binding to pure DNA were subtracted from respective values obtained with nucleoprotein. The average of four replicates of the binding analysis are shown.[201]

pooled as shown in the figure, reannealed to the DNA at varying protein to DNA ratios, and the reconstituted NAP assayed for acceptor activity. All acceptor activity banded in the density region of proteins. Thus, the acceptor factor behaves as a protein with regard to protease substrates and density. The acceptor appears to be a protein(s).

b. Chemical Properties of the Acceptor Proteins

For these studies, the CP-3 fraction containing the acceptor protein was isolated by chromatin-hydroxylapatite chromatography. Briefly, histones (CP-1) are removed by 3 *M* NaCl (pH 6.0). The CP-2 fraction (containing the masking activity) is removed with 4 *M* GuHCl (pH 6.0). The CP-3 fraction (containing the acceptor activity) is then removed with 7 *M* GuHCl (pH 6.0). The CP-3 fraction is then concentrated in an Amicon® hollow fiber dialyzer/concentrator, dialyzed against water, lyophilized, and stored as a lyophilized powder. It is this stock CP-3 fraction that was used in the following analysis and those in Figures 20 through 22.

Figure 20 shows the elution of total protein and the acceptor activity from a 5 × 100 cm column of CL-Sepharose® 6B using 6 *M* GuHCl buffered at pH 6.0 as the solute. The fractions were pooled into six large groups (as shown in the figure) and each fraction reannealed to DNA and assayed for receptor binding. Although some of the activity elutes near the void volume, the bulk of the acceptor elutes near the inclusion volume of the column in the molecular weight range of 14,000 and 18,000. When resins with exclusion sizes, e.g., Sephadex® G-200 or agarose 0.5 *M* were used with 6 *M* GuHCl, the activity elutes near or at the void volumes.

Figure 21 shows the isoelectric focusing pattern of the acceptor activity and total protein using the LKB flat bed gel apparatus with Superfine G-60 Sephadex® in 6 *M* urea and ampholines for a 3 to 10 pH range. After focusing the fractions were pooled

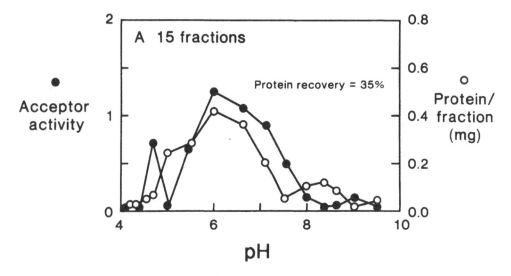

FIGURE 21. Isoelectric focusing of acceptor activity in Sephadex® resin and 6 *M* urea. The CP-3 fraction of chromatin proteins (representing about 50 mg of protein) was applied to isoelectric focusing on flat beds containing ultrafine Sephadex® G-60. The resins contained pH 3-10 ampholines and 6 *M* urea. After 12 hr of focusing, the resin was sectioned, the sections tested for pH and placed in columns, and the protein eluted by 6 *M* GuHCl. The fractions were dialyzed, lyophilized, and the (O) protein quantitated. Portions of the total proteins were reannealed to pure DNA and the (●) acceptor activity (P-R binding) measured.[201]

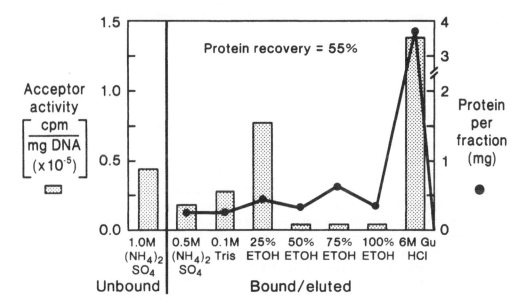

FIGURE 22. Hydrophobic chromatography of the acceptor activity on octyl Sepharose®. The lyophilized CP-3 fraction of chromatin was resuspended in 6 *M* GuHCl (pH 6.0) for several hours. The solution was dialyzed against 1 *M* (NH₂)₂SO₄ and the retentate clarified by centrifugation. The supernatant was applied to the octyl Sepharose® column. Two to three column volumes of each of the solvents listed in the figure were passed through the resin and collected. Each fraction was (●) quantitated for protein and portions of each fraction reconstituted to DNA. The reconstituted NAPs were then tested for P-R binding (acceptor activity), as depicted by the bars.[201]

according to the unit pH and reconstituted to DNA. Two primary peaks with acceptor activity are observed, one focusing around pH 5.0 and the other over a broader range of pH 6 to 7.5. These proteins clearly are not histones since they dissociate from the DNA at much higher GuHCl concentrations than histones and they focus in acidic pH

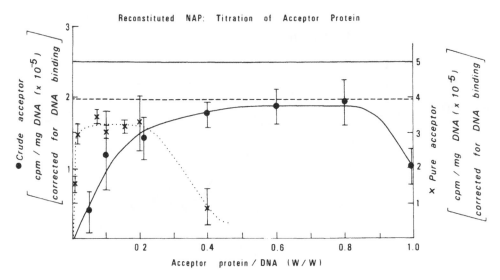

FIGURE 23. Effects of varying quantities of the acceptor protein added to DNA on the acceptor activity of the reconstituted NAP. The acceptor proteins were obtained from the 4 to 7 *M* GuHCl extract from hydroxylapatite-hen oviduct chromatin resin. These were reannealed to the hen DNA in varying quantities using a reverse gradient of 6 *M* GuHCl, to 0 *M* GuHCl, as described in the text. The DNA concentration was 0.2 mg/m*l*. After reconstitution the pelleted DNA-protein was resuspended in the dilute Tris buffer (0.5 mg DNA per m*l*) and analyzed for (●) P-R binding. Pure DNA was also analyzed for hormone binding and the values subtracted from those obtained from the protein-DNA. The (x) represent reconstitution of acceptor proteins purified further by molecular sieve chromatography (using agarose — 1.5 *M*). (——) represents the average binding levels of P-R to native NAP. (From Spelsberg, T. C., Thrall, C. L., Webster, R. A., and Pikler, G. M., *J. Toxicol. and Environ. Health*, 3, 309, 1977. Copyright Hemisphere Publishing Corporation. With permission.)

range while histones focus at the extreme basic pH range (off the scale of the isoelectric focusing plates). The broad range of the one peak of activity on the gel suggests heterogeneity in one species.

That at least two or more species exist is supported in Figure 22 using hydrophobic chromatography where the elution of acceptor activity from octyl Sepharose® with various concentrations of methanol and then 6 *M* GuHCl reveals two major peaks of activity. Two peaks of activity are also detected when other eluting solvents such as ehtylene glycol and chaotropic salts (Na perchlorate, etc.) are used. The major peak of activity consistently elutes only at the higher concentrations of GuHCl or chaotropic salts suggesting the acceptor protein(s) are very hydrophobic. This in turn suggests that the proteins might lie in one of the two grooves of the DNA to interact with the hydrophobic regions (base pairs) in the center of the helix. Alternatively, the proteins might interact with each other in clusters on the DNA.

In any event, the acceptor activity appears to consist of one or more nonhistone proteins bound to specific sequences of the DNA. The proteins appear to be slightly acidic and of low molecular weight. The actual heterogeneity is not presently known since the extent of proteolytic degradation is difficult to determine. Attempts are now in progress to further purify these proteins from bulk quantities of chromatin.

B. Specific Repetitive DNA Sequences Associated with the Acceptor Sites

The 4 to 7 *M* GuHCl fraction from the chromatin hydroxylapatite resin was subsequently used as a source of the acceptor activity and subjected to a series of analyses, first in the reconstitution assays for analysis of the involvement of specific DNA sequences and second as the starting point in the purification and chemical characterization of the acceptor activity.

Figure 23 shows the reconstitution of the acceptor activity (capacity to bind the

DNA SEQUENCE SPECIFICITY FOR ACCEPTOR ACTIVITY

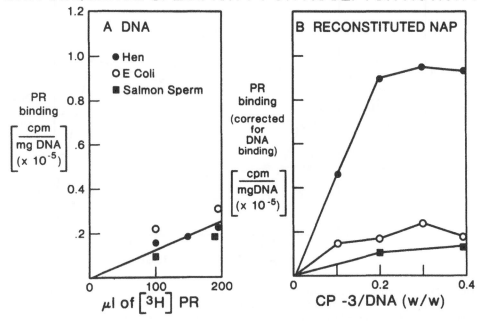

FIGURE 24. Binding of [³H]P-R to reconstituted NAP containing DNA from several sources. Panel A represents the binding of P-R to the pure DNA isolated from different species. The *E. coli* and Salmon sperm DNAs were purchased from PL Biochemicals (Milwaukee, Wis.) and further purified to less than 1% (w/w) protein and RNA. (●) = hen DNA; (○) = *E. coli* DNA; (■) = Salmon sperm DNA. The mean of four replicate analyses of each receptor level for each DNA is presented in each panel. Panel B represents the binding of [³H]P-R to reconstituted NAP containing the different DNAs. The NAPs were reconstituted using the various DNAs and increasing levels of hen oviduct acceptor protein (CP-3 fraction) as described in the legend of Figure 23. The binding assays were performed at a saturating receptor level. The various reconstituted NAPs contained (●) = hen DNA, (■) = Salmon sperm DNA, (○) = *E. coli* DNA. The binding values to the various pure DNA preparations were subtracted from the corresponding DNA reconstituted as NAP to give the "corrected for DNA binding".[200]

progesterone receptor) as a function of the ratio of the CP-3 protein to DNA. Two interesting aspects arise from these studies. First, the reconstitution of the acceptor activity to pure DNA is saturable. Second, the level of progesterone receptor binding at this saturation is approximately the same level measured in native (undissociated) NAP (see broken line in Figure 23). These results suggest a DNA sequence specificity for the acceptor protein which is detectable by the reconstitution methods described earlier.

To further analyze this possibility, the CP-3 fraction was reannealed to DNA from different sources containing different sequences. As shown in Figure 24 (Panel A), little differences in P-R binding are detected between bacterial, Salmon sperm, and hen DNA. However, when the CP-3 is reannealed to these DNAs, only the reconstituted NAP containing hen DNA displays the high level of P-R binding as is found with native NAP (Figure 24, Panel B). The CP-3 bound to Salmon sperm or bacterial DNA show no acceptor activity. Thus, not only does DNA seem to be required for acceptor activity but specific sequences of DNA may be required.

Preliminary studies to characterize the DNA sequences involved in the nuclear acceptor sites for the oviduct progesterone receptor have been initiated. As described in Figure 15, only 10 to 20% of the total nonhistone chromosomal proteins in chromatin remains in the NAP as a complex with DNA. An enriched fraction of the protein-

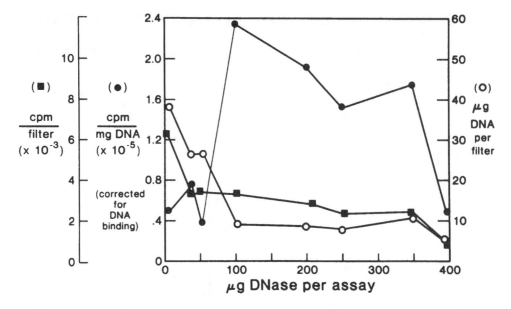

FIGURE 25.   Deoxyribonuclease resistance of the nuclear acceptor sites for P-R in oviduct NAP. Oviduct chromatin was partially deproteinized to obtain nucleoacidic protein (NAD) as described elsewhere.[81] Two milligrams of NAP were treated with deoxyribonuclease (DNase I) for 10 min at 4°C with varying levels of DNase. The assays in 2 m*l* were stopped by the addition of EDTA. This DNase-treated NAP was then analyzed for P-R binding (acceptor activity) using the streptomycin assay.[81] (■) represents P-R binding activity per filter or (●) per milligram DNA and (O) represents the μg DNA per filter recovered from the binding assay. These are plotted against the μg DNase per assay. The mean of four replicate binding assays are shown.[200]

DNA complexes can be achieved by digestion of the NAP with deoxyribonuclease (DNase I). Figure 25 shows the results of deoxyribonuclease (DNase) treatment of NAP on the acceptor activity. Interestingly, as the DNase degrades the bulk of the DNA in the NAP, the acceptor activity is maintained. This results in a marked increase in acceptor activity per unit mass of DNA (bound P-R per milligram DNA). Thus, the acceptor sites are somewhat resistant to the nuclease treatment. Studies with pure DNA indicate that the proteins bound to specific sequences of the DNA in the NAP protect these sequences from the DNAse. Figure 25 also shows that only after extensive DNase treatment is the acceptor activity destroyed.

The DNA in the protein-DNA complexes of the NAP which contain the acceptor activity and which are resistant to the DNAse, were isolated, labeled with [^{32}P] or [^{3}H] by nick translation methods, and used to analyze for sequence complexity by DNA-DNA hybridization. Figure 26 shows that, based on their low cot values from these hybridization studies, these particular DNA fragments are enriched in intermediate to highly repetitive sequences. The exact class of sequences representing those involved in the acceptor sites for the P-R in the avian oviduct remains to be identified.

VII. POTENTIAL FUNCTION(S) OF THE ACCEPTOR PROTEINS AND THE ACCEPTOR DNA SEQUENCES

Figure 27 presents a hypothetical model of the functioning of the acceptor protein(s) and their DNA sequence(s). The model shows one of the two known species of the steroid receptor (the B species) binding to the acceptor protein in the chromatin acceptor sites with the A species binding to the adjacent DNA region to somehow activate transcription. This two subunit receptor theory was devised somewhat empirically based on the nuclear binding properties of the two receptor species.[37,41,42] It should be

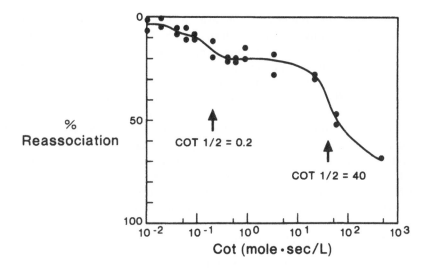

FIGURE 26. Repetitive sequence analysis of DNase resistant DNA in oviduct NAP: reassociation of DNase-resistant labeled DNA with whole hen DNA. The NAP was partially digested with DNase I at a ratio of NAP/DNase (w/w) of 50 as described in the legend of Figure 25. About 15% of the DNA in this NAP is rendered acid soluble by this treatment. The residual DNA from the DNase NAP was isolated by digestion for 2 hr at 37°C with pronase at a level of pronase/NAP of 0.1 (w/w). The preparation was extracted with an equal volume of chloroform-isoamyl alcohol (24:1 w/w). The DNA in the aqueous phase was precipitated with two volumes of ethanol at −20°C. The precipitate was dissolved in dilute Tris buffer. This purified DNA was labeled by nick translation by the method of Rigby et al.[197] and hybridized to whole hen DNA by the method of Weintraub and Groudine.[198] The cot value was normalized by the factor of relative reassociation rate.[199,200]

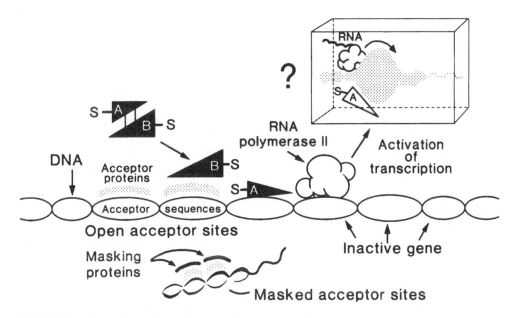

FIGURE 27. Model of possible mechanism of action of steroid receptor complexes on transcription. In this scheme, the steroid-receptor complex enters the nucleus and binds to the acceptor sites, either directly to the acceptor proteins or to the adjacent DNA or both. The receptor would then activate transcription by modulating the DNA structure or affecting the polymerase itself. The acceptor proteins in this instance would serve either as a direct binding site for the steroid receptor or would perturb the DNA structure to allow the steroid-receptor complex to bind to the DNA. In this model, one of the two subunits of the P-R (the B species) binds to the acceptor protein, the A species binds to the adjacent DNA to activate transcription. The lower figure depicts "marked" acceptor sites and the upper figure, the unwinding of the DNA to initiate transcription.

mentioned, however, that the exact mechanism of binding of the receptors to the nuclear acceptor sites remains to be determined. The biological function of the A and B subunits of the chick oviduct progesterone receptor are still unclear.

It is this author's opinion that steroid receptors represent one class of intracellular gene regulators in-eukaryotes. The nuclear acceptor sites may reside in or near structural genes or more likely at a distance at regulatory genes regions. The acceptor proteins and their specific DNA sequences may function only as recognition (binding) sites or may be involved in the regulation of transcription of DNA. Since there are thousands of nuclear acceptor sites, the DNA component may be a class of repetitive sequences (intermediate or highly repetitive), residing at many locations throughout the genome. These sequences or their neighboring sequences could transcribe products such as RNA which in turn regulate these structural genes residing in many locations throughout the genome. These sequences may not necessarily be closed to the steroid-regulated structural genes, but they or their neighboring sequences could transcribe products which in turn regulate these structural genes.

VIII. CONCLUSIONS

This chapter has dealt with the problems and progress in the chemical characterization of the nuclear acceptor sites for steroid receptors with major emphasis on the progesterone receptor of the avian oviduct. These nuclear acceptor sites and their components are important for the understanding of steroid-induced changes in gene transcription and for understanding how genes are regulated. As mentioned in Section V.C, the high affinity binding to chromatin in cell-free assays, the masking phenomenon, and the high affinity saturable binding to the unmasked sites on NAP have been reported for other steroid target tissue systems.[80-83] It is not even known whether the steroid-receptor complex binds directly to the acceptor protein(s) or to the DNA in the vicinity of the acceptor protein-DNA sequence. The stoichiometric relationship between the number of receptors and the number of acceptor proteins or acceptor DNA sequences is also not known. Lastly, if the receptor binds to the DNA, the mechanism by which the acceptor proteins function to enhance this binding is unknown. These and many other questions must first await the elucidation of the exact chemical nature of the acceptor sites, the isolation of the components, and the reconstruction of the acceptor sites with these components. Major effort is presently underway to purify to homogeneity the acceptor proteins for the progesterone receptor of the avian oviduct. It is planned to prepare antisera (whole animal and monoclonal) against these proteins, and to use these antibodies to examine the masked and unmasked acceptor sites, the tissue and species specificity of the acceptor proteins, and to assist in the isolation of the specific DNA sequences bound by acceptor proteins.

Most studies on the quantitative nuclear binding of steroids estimate between 1,000 and 10,000 biologically important nuclear nuclear acceptor sites per cell for a variety of steroid-target tissue systems (see Section IV.D). If the masked sites are the same as the unmasked sites in the hen oviduct chromatin, then about 25,000 acceptor sites per cell are estimated (Table 4). Assuming that there is one P-R per one acceptor protein, that the acceptor proteins average about 15,000 in molecular weight, and using the facts that avian DNA contains 2.5×10^{-12} g DNA per diploid cell and that 1 mg chromatin DNA can be isolated from 1 g of oviduct, it can be calculated that 10 kg of oviduct should yield 2.5 mg acceptor protein at 100% recovery. If the above assumptions are close to the actual state, the purification of the acceptor protein(s) is feasible.

ACKNOWLEDGMENT

The data published here were supported in part by a grant from the NICHD — HD 9140 and the Mayo Foundation.

REFERENCES

1. Spelsberg, T. C., Steggles, A. W., and O'Malley, B. W., Progesterone-binding components of chick oviduct. III. Chromatin acceptor sites, *J. Biol. Chem.*, 246, 4188, 1971.
2. Gorski, J. and Gannon, F., Current models of steroid hormone action: a critique, *Annu. Rev. Physiol.*, 39, 425, 1976.
3. Jensen, E. V. and DeSombre, E. R., Mechanism of action of the female sex hormones, *Annu. Rev. Biochem.*, 41, 203, 1972.
4. O'Malley, B. W. and Means, A. R., Female steroid hormones and target cell nuclei. The effects of steroid hormones on target cell nuclei are of major importance in the induction of new cell functions, *Science*, 183, 610, 1974.
5. Spelsberg, T. C. and Toft, D. O., The mechanism of action of progesterone, in *Receptors and Mechanism of Action of Steroid Hormones*, Part 1, Pasqualini, J. R., Ed., Marcel Dekker, New York, 1976, 261.
6. Thrall, C., Webster, R. A., and Spelsberg, T. C., Receptor interaction with chromatin in *The Cell Nucleus*, Vol. 6, Part 1, Busch, H., Ed., Academic Press, New York, 1978, 461.
7. Harrison, R. W., Fairfield, S., and Orth, D. N., Evidence for glucocorticoid transport through the target cell membrane, *Biochem. Biophys. Res. Commun.*, 61, 1262, 1974.
8. Milgrom, E., Atger, M., and Baulieu, E.-E., Progesterone in uterus and plasma. IV. Progesterone receptor(s) in guinea pig uterus cytosol, *Steroids*, 167, 741, 1970.
9. Peck, E. J., Burgner, J., and Clark, J. H., Estrophile binding sites of the uterus. Relation to uptake and retention of estradiol in vitro, *Biochemistry*, 12, 4596, 1973.
10. Williams, D. and Gorski, J., Equilibrium binding of estradiol by uterine cell suspensions, *Biochemistry*, 13, 5537, 1974.
11. Jensen, E. V. and Jacobson, H. I., Basic guides to the mechanism of estrogen action, *Recent Prog. Horm. Res.*, 18, 387, 1962.
12. Jensen, E. V., Suzuki, T., Kawashima, T., Stumpf, W. E., Jungblut, P. W., and DeSombre, E. R., A two-step mechanism for the interaction of estradiol with rat uterus, *Proc. Natl. Acad. Sci. U.S.A.*, 59, 632, 1968.
13. Stumpf, W. E. and Sar, M., Autoradiographic localization of estrogen, androgen, progestin, and glucocorticosteroid in "target tissues" and "nontarget tissues", in *Receptors and Mechanism of Action of Steroid Hormones*, Part 1, Pasqualini, J. R., Ed., Marcel Dekker, New York, 1976, 41.
14. Talwar, G. P., Segal, J. J., Evans, A., and Davidson, O. W., The binding of estradiol in the uterus: a mechanism for derepression of RNA synthesis, *Proc. Natl. Acad. Sci. U.S.A.*, 52, 1059, 1964.
15. Toft, D. O. and Gorski, J., A receptor molecule for estrogens: isolation from the rat uterus and preliminary characterization, *Proc. Natl. Acad. Sci. U.S.A.*, 55, 1574, 1966.
16. Beato, M., Biesewig, D., Braendle, W., and Sekeris, C. E., On the mechanism of hormone action XV. Subcellular distribution and binding of [1,2-^{3}H]cortisol in rat liver, *Biochim. Biophys. Acta*, 192, 494, 1969.
17. Baulieu, E.-E., Alberga, A., and Jung, I., Recepteurs hormonaux. Liaison de divers oestrogenes a des proteines uterines. Une methode de dosage de l'oestradiol, *C. R. Acad. Sci.*, D265, 501, 1967.
18. Gorski, J., Toft, D. O., Shyamala, D., Smith, A., and Notides, A., Hormone receptors: studies on the interaction of estrogen with the uterus, *Recent Prog. Horm. Res.*, 24, 45, 1968.
19. Little, M., Rosenfeld, G. C., and Jungblut, P. W., Cytoplasmic estradiol "receptors" associated with the "microsomal" fraction of pig uterus, *Hoppe-Seyler's Z. Physiol. Chem.*, 53, 231, 1972.
20. Little, M., Szendro, P. I., and Jungblut, P. W., Hormone-mediated dimerization of microsomal estradiol receptor, *Hoppe-Seyler's Z. Physiol. Chem.*, 354, 1599, 1973.
21. Sheridan, P. J., Buchannon, J. M., Anselmo, V. C., and Martin, P. M., Equilibrium: the intracellular distribution of steroid receptors, *Nature (London)*, 282, 579, 1979.
22. Martin, P. M., and Sheridan, P. J., Intracellular distribution of estrogen receptors: a function of preparation, *Experientia*, 36, 620, 1980.
23. Yamamoto, K. R., Characterization of the 4S to 5S forms of the estradiol receptor protein and their interaction with deoxyribonucleic acid, *J. Biol. Chem.*, 249, 7068, 1974.
24. Notides, A. C. and Nielsen, S., The molecular mechanism of the in vitro 4S to 5S transformation of the uterine estrogen receptor, *J. Biol. Chem.*, 249, 1866, 1974.
25. Cake, M. H., Goidl, J. A., Parchman, L. G., and Litwack, G., Involvement of a low molecular weight component(s) in the mechanism of action of the glucocorticoid receptor, *Biochem. Biophys. Res. Commun.*, 71, 45, 1975.
26. Thrower, S., Hall, C., Lim, L., and Davidson, A. N., The selective isolation of the uterine oestradiol-receptor complex by binding to oligo(dT)-cellulose. The mediation of an essential activator in the transformation of cytosol receptor, *Biochem. J.*, 160, 271, 1976.
27. Puca, G. A., Nola, E., Sica, V., and Bresciani, F., Estrogen binding proteins of calf uterus. Molecular and functional characterization of the receptor transforming factor: a Ca^{2+}-activated protease, *J. Biol. Chem.*, 252, 1358, 1977.

28. **Milgrom, E., Atger, M., and Baulieu, E.-E.**, Acidophilic activation of steroid hormone receptors, *Biochemistry*, 12, 5198, 1973.

29. **Chamness, G. C., Jennings, A. W., and McGuire, W. L.**, Estrogen receptor bindings to isolated nuclei. A nonsaturable process, *Biochemistry*, 13, 327, 1974.

30. **Andre, J. and Rochefort, H.**, In vitro binding of the estrogen receptor to DNA: absence of saturation at equilibrium, *FEBS Lett.*, 50, 319, 1975.

31. **Nishigori, H. and Toft, D.**, Chemical modification of the avian progesterone receptor by pyridoxal phosphate, *J. Biol. Chem.*, 254, 9155, 1979.

32. **Simons, S. S., Martinez, H. M., Gracea, R. L., Baxter, J. D., and Tomkins, G. M.**, Interactions of glucocorticoid receptor-steroid complexes with acceptor sites, *J. Biol. Chem.*, 251, 334, 1976.

33. **Goidl, J. A., Cake, M. H., Dolan, K. P., Parchman, L. G., and Litwack, G.**, Activation of the rat liver glucocorticoid-receptor complex, *Biochemistry*, 16, 2125, 1977.

34. **Bailly, A., Sallas, N., and Milgrom, E.**, A low molecular weight inhibitor of steroid receptor activation, *J. Biol. Chem.*, 252, 858, 1977.

35. **Sherman, M. R., Corvol, P. L., and O'Malley, B. W.**, Progesterone-binding components of chick oviduct. I. Preliminary characterization of cytoplasmic components, *J. Biol. Chem.*, 245, 6085, 1970.

36. **Schrader, W. T. and O'Malley, B. W.**, Progesterone-binding components of chick oviduct. IV. Characterization of purified subunits, *J. Biol. Chem,*. 247, 51, 1972.

37. **Schrader, W. T., Toft, D. O., and O'Malley, B. W.**, Progesterone-binding protein of chick oviduct. VI. Interaction of purified progesterone-receptor components with nuclear constituents, *J. Biol. Chem.*, 247, 2401, 1972.

38. **Boyd, P. A. and Spelsberg, T. C.**, Analysis of the molecular species of the chick oviduct progesterone receptor using isoelectric focusing, *Biochemistry*, 18, 3679, 1979.

39. **Boyd, P. A. and Spelsberg, T. C.**, Seasonal changes in the molecular species and nuclear binding of the chick oviduct progesterone receptor, *Biochemistry*, 18, 3685, 1979.

40. **Schrader, W. T., Heuer, S. S., and O'Malley, B. W.**, Progesterone receptors of chick oviduct: Identification of 6S receptor dimers, *Biol. Reprod.*, 12, 134, 1975.

41. **Buller, R. E., Schwartz, R. J., Schrader, W. T., and O'Malley, B. W.**, Progesterone binding components of chick oviduct. XII. In vitro effect of receptor subunits on gene transcription in vitro, *J. Biol. Chem.*, 251, 5178, 1976.

42. **Vedeckis, W. V., Schrader, W. T., and O'Malley, B. W.**, The chick oviduct progesterone receptor, in *Biochemical Actions of Hormones*, Vol. 5, Litwack, G., Ed., Academic Press, New York, 1978, 321.

43. **Mueller, G. C., Herranen, A. M., and Jervell, K. J.**, Studies on the mechanism of action of estrogens, *Recent Prog. Horm. Res.*, 14, 95, 1958.

44. **Hamilton, T. H., Teng, C. S., and Means, A. R.**, Early estrogen action: nuclear synthesis and accumulation of protein correlated with enhancement of two DNA-dependent RNA polymerase activities, *Proc. Natl. Acad. Sci. U.S.A.*, 59, 1265, 1968.

45. **Knowler, J. T. and Smellie, R. M. S.**, The synthesis of ribonucleic acid in immature rat uterus responding to oestradiol-17β, *Biochem. J.*, 125, 605, 1971.

46. **Chan, L., Means, A. R., and O'Malley, B. W.**, Rates of induction of specific translatable messenger RNAs for ovalbumin and avidin by steroid hormones, *Proc. Natl. Acad. Sci. U.S.A.* 70, 1870, 1973.

47. **Chan, L., Jackson, R. L., O'Malley, B. W., and Means, A. R.**, Synthesis of very low density lipoproteins in the cockeral. Effects of estrogen, *J. Clin. Invest*, 58, 368, 1976.

48. **Harris, S. E., Rosen, J. M., Means, A. R., and O'Malley, B. W.**, Use of a specific probe for ovalbumin messenger RNA to quantitate estrogen-induced gene transcript, *Biochemistry*, 14, 2072, 1975.

49. **Palmiter, R. D. Moore, P. B., Mulvihil, E. R., and Emtage, S.**, A significant lag in the induction of ovalbumin messenger RNA by steroid hormones: a receptor translocation hypothesis, *Cell*, 8, 557, 1976.

50. **Spelsberg, T. C. and Cox, R. F.**, Effects of estrogen and progresterone on transcription, chromatin and ovalbumin gene expression in the chick oviduct, *Biochim. Biophys. Acta*, 435, 376, 1976.

51. **Church, R. B. and McCarthy, B. J.**, Unstable nuclear RNA synthesis following estrogen stimulation, *Biochim. Biophys. Acta*, 199, 103, 1970.

52. **Clark, J. H., Peck, E. J., and Anderson, J. N.**, Oestrogen receptors and antagonism of steroid hormone action, *Nature (London)*, 251, 446, 1974.

53. **Capony, F. and Rochefort, H.** In vivo effect of anti-estrogens on the localization and replenishment of estrogen receptors, *Mol. Cell. Endocrinol.*, 3, 233, 1975.

54. **Jackson, V. and Chalkley, R.**, The binding of estradiol-17β and the bovine endometrial nuclear membrane, *J. Biol. Chem.*, 249, 1615, 1974.

55. **Jackson, V. and Chalkley, R.**, The cytoplasmic estradiol receptors of bovine uterus. Their occurrence, interconversion, and binding properties, *J. Biol. Chem.*, 249, 59, 1974.

56. **Sekeris, C. E. and Lang, N.**, Bindung von [³H]cortison an histone aus rattenleber, *Hoppe-Seyler's Z. Physiol. Chem.*, 340, 92, 1965.

57. **Sluyser, M.**, Binding of hydrocortisone to rat liver histones, *J. Mol. Biol.*, 19, 591, 1966.

58. Sluyser, M., Interaction of steroid hormones with histones in vitro, *Biochim. Biophys. Acta,* 182, 235, 1969.
59. King, R. J. B. and Gordon, J., The association of [6,7-³H]-oestradiol with a nuclear protein, *J. Endocrinol.,* 39, 533, 1967.
60. Puca, G. A., Sica, V., and Nola, E., Identification of a high affinity nuclear acceptor site for estrogen receptor of calf uterus, *Proc. Natl. Acad. Sci. U.S.A.,* 71, 979, 1974.
61. Puca, G. A., Nola, E., Hibner, U., Cicala, G., and Sica, W., Interaction of the estradiol receptor from calf uterus with its nuclear acceptor sites, *J. Biol. Chem.,* 250, 6452, 1975.
62. Mainwaring, W. I. P., Symes, E. K., and Higgins, S. J., Nuclear components responsible for the retention of steroid-receptor complexes especially from the standpoint of the specificity of hormone responses, *Biochem. J.,* 156, 129, 1976.
63. Alberga, A., Jung, I., Massol, N., Raynaud, J.-P., Raynaud-Jammet, C., Rochefort, H., Truong, H., and Baulieu, E.-E., Estradiol receptors in the uterus, in *Advances in the Biosciences,* Vol. 7, Schering Workshop on Steroid Hormone Receptors, Raspe, G., Ed., Pergamon Press, New York, 1971, 45.
64. Spelsberg, T. C., Steggles, A. W., Chytil, F., and O'Malley, B. W., Progesterone-binding components of chick oviduct. V. Exchange of progesterone-binding capacity from target to nontarget tissue chromatins, *J. Biol. Chem.,* 247, 1368, 1972.
65. Spelsberg, T. C., Webster, R. A., and Pikler, G. M., Multiple binding sites for progesterone in the hen oviduct nucleus: evidence that acidic proteins represent the acceptors, in *Chromosomal Proteins and Their Role in Gene Expression,* Stein, G. and Kleinsmith, L., Eds., Academic Press, New York, 1975, 153.
66. Spelsberg, T. C., Webster, R. A., and Pikler, G. M., Chromosomal proteins regulate steroid binding to chromatin, *Nature (London),* 262, 65, 1976.
67. Spelsberg, T. C., Webster, R., Pikler, G., Thrall, C., and Wells, D., Role of nuclear proteins as high affinity sites ("acceptors") for progesterone in the avian oviduct, *J. Steroid Biochem.,* 7, 1091, 1976.
68. Spelsberg, T. C., Webster, R., Pikler, G., Thrall, C., and Wells, D., Nuclear binding sites (acceptors) for progesterone in avian oviduct: characterization of the highest affinity sites, *Ann. N.Y. Acad. Sci.,* 286, 43, 1977.
69. Spelsberg, T. C., Thrall, C. L., Webster, R. A., and Pikler, G. M., Isolation and characterization of the nuclear acceptor that binds the progesterone-receptor complex in hen oviduct, *J. Toxicol. Environ. Health,* 3, 309, 1977.
70. Spelsberg, T. C., Knowler, J., Boyd, P. A., Thrall, C. L., and Martin-Dani, G., Support for chromatin acidic proteins as acceptors for progesterone in the chick oviduct, *J. Steroid Biochem,.* 11, 373, 1979.
71. Tymoczko, J. L. and Liao, S., Retention of an androgen-protein complex by nuclear chromatin aggregates: heat-labile factors, *Biochim. Biophys. Acta,* 252, 607, 1971.
72. Baxter, J. D., Rousseau, G. G., Bensen, M. C., Garcea, R. L., Ito, J., and Tomkins, G. M., Role of DNA and specific cytoplasmic receptors in glucocorticoid action, *Proc. Natl. Acad. Sci. U.S.A.,* 69, 1892, 1972.
73. King, R. J. B. and Gordon, J., Involvement of DNA in the acceptor mechanism for uterine oestradiol receptor, *Nature New Biol.,* 240, 185, 1972.
74. Liang, T. and Liao, S., Interaction of estradiol- and progesterone-receptors with nucleoprotein: heat-labile acceptor factors, *Biochim. Biophys. Acta,* 277, 590, 1972.
75. Liao, S., Liang, T., and Tymoczko, J. L., Structural recognitions in the interactions of androgens and receptor proteins and in their association with nuclear acceptor components, *J. Steroid Biochem.,* 3, 401, 1972.
76. O'Malley, B. W., Spelsberg, T. C., Schrader, W. T., Chytil, F., and Steggles, A. W., Mechanisms of interaction of a hormone-receptor complex with the genome of a eukaryotic target cell, *Nature (London),* 235, 141, 1972.
77. Lebeau, M. C., Maisol, N., and Baulieu, E.-E., An insoluble receptor for oestrogens in the "residual" nuclear proteins of chick liver, *Eur. J. Biochem.,* 36, 294, 1973.
78. Defer, N., Dastugue, B., and Kruh, J., Direct binding of corticosterone in estradiol to rat liver nuclear non-histone proteins, *Biochimie,* 56, 559, 1974.
79. Gschwendt, M., Solubilization of the chromatin-bound estrogen receptor from chicken liver and fractionation on hydroxylapatite, *Eur. J. Biochem.,* 67, 411, 1976.
80. Klyzsejko-Stefanowicz, L., Chui, J. F., Tsai, Y. H., and Hnilica, L. S., Acceptor proteins in rat androgenic tissue chromatin, *Proc. Natl. Acad. Sci. U.S.A.,* 73, 1954, 1976.
81. Webster, R. A., Pikler, G. M., and Spelsberg, T. C., Nuclear binding of progesterone in the chick oviduct: multiple binding sites in vivo and transcriptional response, *Biochem. J.,* 156, 409, 1976.
82. Perry, B. N. and Lopez, A., The binding of ³H-labelled oestradiol- and progesterone-receptor complexes to hypothalamic chromatin of male and female sheep, *Biochem. J.,* 176, 873, 1978.

83. **Hamana, K. and Iwai, K.**, Glucocorticoid-receptor complex binds to nonhistone protein and DNA in rat liver chromatin, *J. Biochem. (Tokyo)*, 83, 279, 1978.

84. **Harris, C. S. and Shyamala, G.**, Studies on the nature of estrogen specific binding sites in the nuclei, *Nature New Biol.*, 231, 246, 1971.

85. **Clemens, L. E. and Kleinsmith, L. J.**, Specific binding of the oestradiol-receptor complex to DNA, *Nature New Biol.*, 237, 204, 1972.

86. **Musliner, T. A. and Chader, G. J.**, Estradiol-receptors of the rat uterus: interaction of the cytoplasmic estrogen-receptor with DNA in vitro, *Biochim. Biophys. Acta*, 262, 256, 1972.

87. **Toft, D. O.**, The interaction of uterine estrogen receptors with DNA, *J. Steroid Biochem.*, 3, 515, 1972.

88. **Yamamoto, K. R. and Alberts, B. M.**, In vitro conversion of estradiol-receptor protein to its nuclear form: dependence on hormones and DNA, *Proc. Natl. Acad. Sci. U.S.A.*, 69, 2105, 1972.

89. **Yamamoto, K. R. and Alberts, B. M.**, On the specificity of the binding of the estradiol receptor protein to deoxyribonucleic acid, *J. Biol. Chem.*, 249, 7076, 1974.

90. **Yamamoto, K. R. and Alberts, B. M.**, The interaction of estradiol-receptor protein with the genome: an argument for the existence of undetected specific sites, *Cell*, 4, 301, 1975.

91. **Yamamoto, K. R. and Alberts, B. M.**, Steroid receptors: elements of modulation of eukaryotic transcription, *Annu. Rev. Biochem.*, 45, 722, 1976.

92. **Higgins, S. J., Rousseau, G. G., Baxter, J. D., and Tomkins, G. M.**, Nuclear binding of steroid receptors: comparison in intact cells and cell-free systems, *Proc. Natl. Acad. Sci. U.S.A.*, 70, 3415, 1973.

93. **Higgins, S. J. Rousseau, G. G., Baxter, J. D., and Tomkins, G. M.**, Early events in glucocorticoid action activation of the steroid receptor and its subsequent specific nuclear binding studied in a cell-free system, *J. Biol. Chem.*, 248, 5866, 1973.

94. **Higgins, S. J., Rousseau, G. G., Baxter, D. D., and Tomkins, G. M.**, Nature of nuclear acceptor sites for glucocorticoid- and estrogen-receptor complexes, *J. Biol. Chem.*, 248, 5873, 1973.

95. **Yamamoto, K. R., Stampfer, M. R., and Tomkins, G. M.**, Receptors from glucocorticoid-sensitive lymphoma cell and two classes of insensitive clones: physical and DNA binding properties, *Proc. Natl. Acad. Sci. U.S.A.*, 71, 3901, 1974.

96. **Rousseau, G. G., Higgins, S., Baxter, J. D., Gelfand, D., and Tomkins, G. M.**, Binding of glucocorticoid receptors in DNA, *J. Biol. Chem.*, 250, 6015, 1975.

97. **Liao, S., Liang, T., and Tymoczko, J. L.**, Ribonucleoprotein binding of steroid-"receptor" complexes, *Nature New Biol.*, 241, 211, 1973.

98. **Barrack, E. R. and Coffey, D. S.**, The specific binding of estrogens and androgens to the nuclear matrix of sex hormone responsive tissues, *J. Biol. Chem.*, 255, 7265, 1980.

99. **Barrack, E. R., Hawkins, E. F., Allen, S. L., Hicks, L. L., and Coffey, D. S.**, Concepts related to salt resistant estradiol receptors in rat uterine nuclei: nuclear matrix, *Biochem. Biophys. Res. Commun.*, 79, 829, 1977.

100. **Alberta, A., Ferrez, M., and Baulieu, E.-E.**, Estradiol-receptor-DNA interaction: liquid polymer phase partition, *FEBS Lett.*, 61, 223, 1976.

101. **Bugany, H. and Beato, M.**, Binding of the partially purified glucocorticoid receptor of rat liver to chromatin and DNA, *Mol. Cell. Endocrinol.*, 7, 49, 1977.

102. **Simons, S. S.** Glucocorticoid receptor-steroid complex binding to DNA competition between DNA and DNA-cellulose, *Biochim. Biophys. Acta*, 496, 349, 1977.

103. **Kallos, J. and Hollander, V.**, Assessment of specificity of oestrogen receptor-DNA interaction by a competitive assay, *Nature (London)*, 272, 177, 1978.

104. **Cidlowski, J. A. and Munck, A.**, Comparison of glucocorticoid-receptor complex binding to nuclei and DNA cellulose, *Biochim. Biophys. Acta*, 543, 545, 1978.

105. **Thanki, K., Beach, T., and Dickerman, H.**, Selective binding of mouse estradiol-receptor complexes to oligo(dT)-cellulose, *J. Biol. Chem.*, 253, 7744, 1978.

106. **Kallos, J., Fasy, T., Hollander, V., and Beck, M.**, Estrogen receptor has enhanced affinity for bromodeoxyuridine-substituted DNA, *Proc. Natl. Acad. Sci. U.S.A.*, 75, 4896, 1978.

107. **Thrall, T. C. and Spelsberg, T. C.**, Factors affecting the binding of chick oviduct progesterone receptor to DNA: evidence that DNA alone is not the nuclear acceptor site, *Biochemistry*, 19, 4130, 1980.

108. **Toft, D. O.**, The interaction of uterine estrogen receptor with DNA, in *Advances in Experimental Medicine and Biology*, Vol. 36, O'Malley, B. W. and Means, A. R., Eds., Plenum Press, New York, 1973, 85.

109. **Socher, S. H., Krall, J. F., Jaffe, R. C., and O'Malley, B. W.**, Distribution of binding sites for the progesterone receptor within chick oviduct chromatin, *Endocrinology*, 99, 891, 1976.

110. **Yamamoto, K. R., Gehring, U., Stampfer, M. R., and Sibley, C. H.**, Genetic approaches to steroid hormone action, *Recent Prog. Horm. Res.*, 32, 3, 1976.

111. **Spelsberg, T. C., Pikler, G. M., and Webster, R. A.**, Progesterone binding to hen oviduct genome: specific versus nonspecific binding, *Science*, 194, 197, 1976.

112. Pikler, G. M., Webster, R. A., and Spelsberg, T. C., Nuclear binding of progesterone in hen oviduct. Binding to multiple sites in vitro, *Biochem. J.*, 156, 399, 1976.

113. Andre, J., Pfeiffer, A., and Rochefort, H., Inhibition of estrogen-receptor-DNA interaction by intercalating drugs, *Biochemistry*, 15, 2964, 1976.

114. Buller, R. E., Toft, D. O., Schrader, W. T., and O'Malley, B. W., Progesterone-binding components of chick oviduct. VIII. Receptor activation and hormone-dependent binding to purified nuclei, *J. Biol. Chem.*, 250, 801, 1975.

115. Buller, R. E. and O'Malley, B.W., Biology and mechanism of steroid hormone receptor interaction with eucaryotic nucleus: a review, *Biochem. Pharmacol.*, 25, 1, 1976.

116. Spelsberg, T. C., Thrall, C. L., Martin-Dani, G., Webster, R. A., and Boyd, P. A., Steroid receptor interaction with chromatin, in *Ontogeny of Receptor and Reproductive Hormone Action*, Hamilton, T. H., Clark, J. H., and Sadler, W. A., Eds., Raven Press, New York, 1979, 31.

117. Spelsberg, T. C. and Halberg, F., Circannual rhythms in steroid receptor concentration and nuclear binding in chick oviduct, *Endocrinology*, 107, 1234, 1980.

118. King, R. J. B., Gordon, J., and Inman, D. R., The intracellular localization of oestrogen in rat tissues, *J. Endocrinol.*, 32, 9, 1965.

119. King, R. J. B., Gordon, J., and Martin, L., The association of [6,7-^{3}H]oestradiol with nuclear chromatin, *Biochem. J.*, 97, 28P, 1965.

120. Swaneck, G. E., Chu, L. L. H., and Edelman, I. S., Stereoscopic binding of aldosterone and renal chromatin, *J. Biol. Chem.*, 245, 5282, 1970.

121. Spelsberg, T. C., The role of nuclear acidic proteins in binding steroid hormones, in *Acidic Proteins of the Nucleus*, Cameron, I. L. and Jeter, J. R., Jr., Eds., Academic Press, New York, 1974, 249.

122. Fanestil, D. D. and Edelman, I. S., Characteristics of the renal nuclear receptors for aldosterone, *Proc. Natl. Acad. Sci. U.S.A.*, 56, 872, 1966.

123. Wagner, T. E., A trypsin sensitive site for the action of hydrocortisone on calf thymus nuclei, *Biochem. Biophys. Res. Commun.*, 38, 890, 1970.

124. Middlebrook, J. L., Wong, M. D., Ishii, D. N., and Aronow, L., Subcellular distribution of glucocorticoid receptors in mouse fibroblasts, *Biochemistry*, 14, 180, 1975.

125. Shyamala, G., Estradiol receptors in mouse mammary tumors: absence of the transfer of bound estradiol from the cytoplasm to the nucleus, *Biochem. Biophys. Res. Commun.*, 46, 1623, 1972.

126. Lohmar, P. H. and Toft, D. O. Inhibition of the binding of progesterone receptor to nuclei: effects of o-phenanthroline and rifamycin AF/013, *Biochem. Biophys. Res. Commun.*, 67, 8, 1975.

127. Webster, R. A. and Spelsberg, T. C., Steroid receptor binding to nuclei: effect of assay conditions on the integrity of chromatin, *J. Steroid Biochem.*, 10, 343, 1979.

128. Dounce, A. L. and Umana, R., The proteases of isolated cell nuclei, *Biochemistry*, 1, 811, 1962.

129. Furlan, M. and Jericijo, M., Protein catabolism in thymus nuclei. II. Binding of histone-splitting nuclear proteases to deoxyribonucleic acid, *Biochim. Biophys. Acta*, 147, 145, 1967.

130. Panyim, S., Jensen, R. H., and Chalkley, R., Proteolytic contamination of calf thymus nucleohistone and its inhibition, *Biochim. Biophys. Acta*, 160, 252, 1968.

131. Panyim, S. and Chalkley, R., High resolution acrylamide gel electrophoresis of histones, *Arch. Biochem. Biophys.*, 130, 337, 1969.

132. Spelsberg, T. C., Hnilica, L. S., and Ansevin, A. T., Proteins of chromatin in template restriction. III. The macromolecules in specific restriction of the chromatin DNA, *Biochim. Biophys. Acta*, 228, 550, 1971.

133. Sherman, M. R. and Diaz, S. C., Discussion paper: mero-receptor formation from a larger subcomponent of the oviduct progesterone receptor, *Ann. N. Y. Acad. Sci.*, 286, 81, 1977.

134. Sherman, M. R., Tuazon, F. B., Diaz, S. C., and Miller, L. K., Multiple forms of oviduct progesterone receptors analyzed by ion exchange filtration and gel electrophoresis, *Biochemistry*, 15, 980, 1976.

135. Sherman, M. R., Pickering, L. A., Rollwagen, F. M., and Miller, L. K., Mero-receptors: proteolytic fragments of receptors containing the steroid-binding site, *Fed. Proc.*, 37, 167, 1978.

136. Vedeckis, W. V., Freeman, M. R., Schrader, W. T., and O'Malley, B. W., Progesterone-binding components of chick oviduct: partial purification and characterization of a calcium-activated protease which hydrolyzes the progesterone receptor, *Biochemistry*, 19, 335, 1980.

137. Schrader, W. T., Coty, W. A., Smith, R. G., and O'Malley, B. W., Purification and properties of progesterone receptors from chick oviduct, *Ann. N. Y. Acad. Sci.*, 286, 64, 1977.

138. Schrader, W. T., Kuhn, R. W., and O'Malley, B. W., Progesterone binding components of chick oviduct. XIII. Receptor B subunit protein purified to apparent homogeneity from laying hen oviducts, *J. Biol. Chem.*, 252, 299, 1977.

139. Milgrom, E. and Atger, M., Receptor translocation inhibitor and apparent saturability of the nuclear acceptor, *J. Steroid Biochem.*, 6, 487, 1975.

140. Clark, J. H. and Gorski, J., Estrogen receptors: an evaluation of cytoplasmic-nuclear interactions in a cell-free system and a method for assay, *Biochim. Biophys. Acta*, 192, 508, 1969.

141. Chamness, G. C., Jennings, A.W., and McGuire, W. L., Oestrogen receptor binding is not restricted to target nuclei, *Nature New Biol.,* 241, 458, 1973.
142. Kon, O. L., Webster, R. A., and Spelsberg, T. C., Isolation and characterization of the estrogen receptor in the hen oviduct: evidence for two molecular species, *Endocrinology,* 107, 1182, 1980.
143. Raynaud-Jammet, C. and Baulieu, E.-E., Action de l'oestradiol in vitro: augmentation de la bio-synthese d'acide ribonucleique dans les noyaux uterins, *C. R. Acad. Sci.,* D268, 3211, 1969.
144. Arnaud, M. Beziat, Y., Guilleux, J. C., Hough, A., Hough, D., and Mousseron-Canet, M., Les recepteurs de l'oestradiol dans l'uterus de genisse stimulation de la biosynthese de RNA in vitro, *Biochim. Biophys. Acta,* 232, 117, 1971.
145. Arnaud, M., Beziat, Y., Borgna, J. L., Guilleux, J. C., and Mousseron-Canet, M., Le receptor de l'oestradiol, l'amp cyclique et la RNA polymerase nucleolaire dans l'uterus de genisse. Stimulation de la biosynthesis de RNA in vitro, *Biochim. Biophys. Acta,* 254, 241, 1971.
146. Beziat, Y., Guilleux, J. C., and Mousseron-Canet, M., Effect de l'oestradiol et de ses recepteurs sur la biosynthese du RNA dans les nayaux isoles de l'euterus de genisee, *C. R. Acad. Sci.,* D270, 1620, 1970.
147. Mohla, S., DeSombre, E. R., and Jensen, E. V., Tissue-specific stimulation of RNA synthesis by transformed estradiol-receptor complex, *Biochem. Biophys. Res. Commun.,* 46, 661, 1972.
148. Jensen, E. V., Brecher, P. I., Numata, M., Mohla, S., and DeSombre, E. R., Transformed estrogen receptor in the regulation of RNA synthesis in uterine nuclei, *Adv. Enzyme Regul.,* 11, 1, 1973.
149. Jensen, E. V., Mohla, S., Gorell, T. A., and DeSombre, E. R., The role of estrophilin in estrogen action, *Vitam. Horm.,* 32, 89, 1974.
150. Buller, R. E., Schwartz, R. J., and O'Malley, B.W., Steroid hormone receptor fraction stimulation of RNA synthesis: a caution, *Biochem. Biophys. Res. Commun.,* 69, 106, 1976.
151. Schwartz, R. J., Tsai, M. J., Tsai, S. Y., and O'Malley, B. W., Effect of estrogen on gene expression in the chick oviduct. V. Changes in the number of RNA polymerase binding and initiation sites in chromatin, *J. Biol. Chem.,* 250, 5175, 1975.
152. Schwartz, R. J., Kuhn, R. W., Buller, R. E., Schrader, W. T., and O'Malley, B. W., Progesterone-binding components of chick oviduct. In vitro effects of purified hormone-receptor complexes on the interaction of RNA synthesis in chromatin, *J. Biol. Chem.,* 251, 5166, 1976.
153. Schwartz, R. J., Schrader, W. T., and O'Malley, B. W., Mechanism of steroid hormone action: in vitro control of gene expression in chick oviduct chromatin by purified steroid receptor complexes, in *Juvenile Hormones,* Gilbert, L., Ed., Plenum Press, New York, 1976, 530.
154. Spelsberg, T. C., Nuclear binding of progesterone in the chick oviduct: multiple binding sites in vivo and transcriptional response, *Biochem. J.,* 156, 391, 1976.
155. Chytil, F. and Spelsberg, T. C., Tissue differences in antigenic properties of non-histone protein-DNA complexes, *Nature New Biol.,* 233, 215, 1971.
156. Chytil, F., Immunochemical characteristics of chromosomal proteins, *Methods Enzymol.,* 40, 191, 1975.
157. Greene, G. L., Closs, D. E., Fleming, H., DeSombre, E. R., and Jensen, E. V., Antibodies to estrogen receptor: immunochemical similarity of estrophilin from various mammalian species, *Proc. Natl. Acad. Sci. U.S.A.,* 74, 3681, 1977.
158. Greene, G. L., Nolan, C., Engler, J. P., and Jensen, E. V., Monoclonal antibodies to human estrogen-receptor, *Proc. Natl. Acad. Sci. U.S.A.,* 77, 5115, 1980.
159. Eisen, H. J., An antiserum to the rat liver glucocorticoid receptor, *Proc. Natl. Acad. Sci. U.S.A.,* 77, 3893, 1980.
160. O'Malley, B. W., McGuire, W. L., Kohler, P. O., and Korenman, S. G., Studies on the mechanism of steroid hormone regulation of synthesis of specific proteins, *Recent Prog. Horm. Res.* 25, 105, 1969.
161. Chytil, F., Glasser, S. R., and Spelsberg, T. C., Alterations in liver chromatin during perinatal development of the rat, *Dev. Biol.,* 37, 295, 1974.
162. Brecher, P. I., Vigerski, R., and Wotiz, H. S., An in vitro system for the binding of estradiol to rat uterine nuclei, *Steroids,* 10, 635, 1967.
163. Musliner, T. A., Chader, G. J., and Villee, C. A., Studies on estradiol receptors of the rat uterus. Nuclear uptake in vitro, *Biochemistry,* 9, 4448, 1970.
164. King, R. J. B., Beard, V., Gordon, J., Pooley, A. S., Smith, J. A., Steggles, A. W., and Vertes, M., Studies on estradiol-binding in mammalian tissues, *Adv. Biosci.,* 7, 21, 1971.
165. Mainwaring, W. I. P. and Peterkin, B. M., A reconstituted cell-free system for the specific transfer of steroid-receptor complexes into nuclear chromatin isolated from rat ventral prostate gland, *Biochem. J.,* 125, 285, 1971.
166. O'Malley, B. W., Toft, D. O., and Sherman, M. R., Progesterone-binding components of chick oviduct. II. Nuclear components, *J. Biol. Chem.,* 246, 1117, 1971.
167. Steggles, A. W., Spelsberg, T. C., Glasser, S. R., and O'Malley, B. W., Soluble complexes between steroid hormones and target-tissue receptors bind specifically to target-tissue chromatin, *Proc. Natl. Acad. Sci. U.S.A.,* 68, 1479, 1971.

168. Gschwendt, M. and Hamilton, T.H., The transformation of the cytoplasmic oestradiol-receptor complex into the nuclear complex in a uterine cell-free system, *Biochem. J.*, 128, 611, 1972.

169. Kalimi, M., Beato, M., and Feigelson, P., Interaction of glucocorticoids with rat liver nuclei, I. Role of the cytosol proteins, *Biochemistry*, 12, 1627, 1973.

170. Buller, R. E., Schrader, W. T., and O'Malley, B. W., Progesterone-binding components of chick oviducts. IX. Kinetics of nuclear binding, *J. Biol. Chem.*, 250, 809, 1975.

171. Jaffe, R. C., Socher, S. H., and O'Malley, B. W., An analysis of the binding of the chick oviduct progesterone-receptor to chromatin, *Biochim. Biophys. Acta*, 399, 403, 1975.

172. Saffran, J., Loeser, B. K., Bohnett, S. A., and Faber, L. E., Binding of progesterone receptor by nuclear preparations of rabbit and guinea pig uterus, *J. Biol. Chem.*, 251, 5607, 1976.

173. Tanford, C., Protein denaturation, in *Advances in Protein Chemistry*, Vol. 23, Anfinsen, C. B., Anson, M. L., Edsall, J. T., and Richards, F. M., Eds., Academic Press, New York, 1968, 122.

174. Teipel, J. W. and Kochland, D. E., Kinetic aspects of conformational changes in proteins. I. Rate of regain of enzyme activity from denatured proteins, *Biochemistry*, 10, 792, 1971.

175. Weber, K. and Kuter, D. J., Reversible denaturation of enzymès by sodium dodecyl sulfate, *J. Biol. Chem.*, 246, 4504, 1971.

176. Teipel, J. W., In vitro assembly of aldolase. Kinetics of refolding, subunit reassociation, and reactivation, *Biochemistry*, 11, 4100, 1972.

177. Yazgan, A. and Henkens, R. W., Role of zinc (II) in the refolding of guanidine hydrochloride denatured bovine carbonic anhydrase, *Biochemistry*, 11, 1314, 1972.

178. Carlsson, U., Henderson, L. E., and Lindskog, S., Denaturation and reactivation of human carbonic anhydrases in guanidine hydrochloride and urea, *Biochim. Biophys. Acta*, 310, 367, 1973.

179. Ahmad, F. and Salahuddin, A., Reversible unfolding of major fraction of ovalbumin by guanidine-hydrochloride, *Biochemistry*, 15, 5168, 1976.

180. Lykins, L. F., Akey, C. W., Christian, E. G., Duval, G. E., and Topham, R. W., Dissociation and reconstitution of human ferroxidase II, *Biochemistry*, 16, 693, 1977.

181. Spelsberg, T. C. and Hnilica, L. S., Deoxyribonucleoproteins and the tissue-specific restriction of the deoxyribonucleic acid in chromatin, *Biochem. J.*, 120, 435, 1970.

182. Paul, J., Gilmour, R. S., Affara, N., Birnie, G. D., Harrison, B. P., Hell, A., Humpheries, S., Windass, J., and Young, B., The globin gene: structure and expression, *Cold Spring Harbor Symp. Quant. Biol.*, 38, 885, 1973.

183. Axel, R., Melchior, W., Sollner-Webb, B., and Felsenfeld, G., Specific sites of interaction between histones and DNA in chromatin (nuclease DNA-electrophoresis), *Proc. Natl. Acad. Sci. U.S.A.*, 71, 4101, 1974.

184. Barrett, T., Maryanka, D., Hamlyn, P. H., and Gould, H. J., Nonhistone proteins control gene expression in reconstituted chromatin, *Proc. Natl. Acad. Sci. U.S.A.*, 71, 5057, 1974.

185. Stein, G. S., Spelsberg, T. C., and Kleinsmith, L. J., Nonhistone chromosomal proteins and gene regulation. Nonhistone chromosomal proteins may participate in the specific regulation of gene transcription in eukaryotes, *Science*, 183, 817, 1974.

186. Stein, G. S., Mans, R. J., Gabbay, E. J., Stein, J. L., Davis, J., and Adawadkar, P. D., Evidence for fidelity of chromatin reconstitution, *Biochemistry*, 14, 1859, 1975.

187. Chae, C. B., Reconstitution of chromatin: mode of reassociation of chromosomal proteins, *Biochemistry*, 14, 900, 1975.

188. Gadski, R. A. and Chae, C. B., Mode of reconstitution of chicken erythrocyte and reticulocyte chromatin, *Biochemistry*, 15, 3812, 1976.

189. Woodcock, C. L. F., Reconstitution of chromatin subunits, *Science*, 195, 1350, 1977.

190. Woodcock, C. L. F., Frado, L. L. Y., and Wall, J. S., Composition of native and reconstituted chromatin particles — direct mass determination by scannning-transmission electron microscopy, *Proc. Natl. Acad. Sci. U.S.A.*, 77, 4818, 1980.

191. Raspe, G., Ed., *Advances in Biosciences*, Vol. 7, Pergamon Press, New York, 1971.

192. King, R. J. B. and Mainwaring, W. I. P., Eds., *Steroid Cell Interactions*, University Park Press, Baltimore, Md., 1974.

193. Pasqualini, J. R., Ed., *Receptors and Mechanism of Action of Steroid Hormones*, Parts 1 and 2, Marcel Dekker, New York, 1976.

194. Kuhn, R. W., Schrader, W. T., Coty, W. A., Conn, P. M., and O'Malley, B. W., Progesterone binding components of chick oviduct. XIV. Biochemical characterization of purified oviduct progesterone receptor B subunit, *J. Biol. Chem.*, 252, 308, 1977.

195. Tomkins, G. M. Regulation of specific protein synthesis in eucaryotic cells, *Cold Spring Harbor Symp. Quant. Biol.*, 35, 635, 1970.

196. Clark, J. H., Anderson, J. N., and Peck, E. J., Nuclear receptor-estrogen complexes of rat uteri: concentration-time-response parameters, *Adv. Exp. Biol. Med.*, 36, 15, 1973.

197. Rigby, P. W. J., Dieckmann, M., Rhodes, C., and Berg, P., Labeling DNA to high specific activity in vitro by nick translation with DNA polymerase I, *J. Mol. Biol.*, 113, 237, 1977.

198. Weintraub, H. and Groudine, M., Chromosomal subunits in active genes have an altered conformation: Globin genes are digested by deoxyribonuclease I in red blood cell nuclei but not in fibroblast nuclei, *Science,* 193, 848, 1976.

199. Wetmur, J. G. and Davidson, N., Kinetics of renaturation of DNA, *J. Mol. Biol.,* 31, 349, 1968.

200. Toyoda, H. and Spelsberg, T. C., in preparation.

201. Spelsberg, T. C., unpublished results.

202. Martin-Dani, G. and Spelsberg, T. C., in preparation.

203. Kon, O. L. and Spelsberg, T. C., *Endocrinology,* in press.

204. Boyd, P. A. and Spelsberg, T. C., in preparation.

205. Ruh, T., personal communication.

206. Leavitt, W., personal communication.

207. Spelsberg, T. C. and Toyoda, H., in preparation.

Chapter 4

CHEMICAL CARCINOGENESIS

Michael Gronow

TABLE OF CONTENTS

I. INTRODUCTION

The existence of chemical carcinogens in man's environment has been known for about two centuries. For example, Hill in 1761 discovered the clinical symptoms of cancer in snuff users; Percival Pott in 1775 reported a causal association between scrotal cancer and coal soot. However, experimental work on the subject was only begin in 1915 when two Japanese workers isolated the active cancer-producing materials from coal tar — now known to be polycylic aromatic hydrocarbons.

Since this time many other chemical types of carcinogens have been identified in laboratory animals[1-3] and, by inference, in man. A list of the most important chemical carcinogens is given in Table 1. In some cases a chemical carcinogen has been directly connected with the incidence of a particular tumor in man, for example, as in the case of β-naphthylamine (used in the dye industry) and bladder cancer, In other cases, such as cigarette smoking, the exact etiology is less clear-cut as multiple cancer-producing materials are present.

As a result of such experimental studies on carcinogens over the last 40 years or so, various investigators have pronounced that 50 to 90% of all human cancers are produced by chemicals. Thankfully the figure now shows signs of dropping to the lower estimate! However, despite the multifactoral aspects of some human cancers there is no doubt about the major role of certain chemical classes in the genesis of this disease. Early recognition of this fact has led many researchers to use known chemical carcinogens to study the biochemistry of cancer initiation, promotion, and growth.[4] As a result, much useful information has accumulated in the recent literature, particularly with regard to the binding of carcinogens (usually radiolabeled) to various tissue macromolecules.

The chemical structure of a carcinogen often gives no indication of its carcinogenic potency — often seemingly inert compounds are potent carcinogens, e.g., polycyclic hydrocarbons and nitrosamines. Also there is no rationale relating to the chemical structure and the dose required to produce a tumor; or the site of the tumor produced. However, the formation of tumors from many carcinogens has been shown to be dose dependent. Table 2 lists some of the carcinogens commonly used for biochemical and pathological investigations in animal systems

As a result of work done on carcinogens using the experimental systems (some of which are quoted in Table 2), the Millers in the U.S. first recognized, some 20 years ago, that metabolic activation of a carcinogen occupied a crucial role in determining its potency.[5] Ironically, the animal body often creates an "active" carcinogenic species from an otherwise inert chemical by the very mechanisms designed to protect itself, i.e., by detoxification. The activated metabolites of carcinogens are generally electrophilic reagents — termed "ultimate carcinogens" — which react with the nucleophilic centers of cellular molecules to initiate and promote carcinogenesis. Once formed, the electrophilic reactant can, and does, bind to any number of cellular molecules, some of which may be nonessential, and others which are present in such high concentration that the relatively small amount of inactivation that takes place is of no great consequence to the health of the cell. Nowadays, the most critical targets for attack are considered to be the stable macromolecules of the cell, particularly those of the genetic material. A scheme showing how this interaction may lead to cancer is given in Figure 1.

Since the Millers first demonstrated the firm binding of an azo dye carcinogen to liver proteins, many oncogenic chemicals have been found to interact with the cell proteins of different organs and tissues.[6,7] Subsequent studies have confirmed the binding of a wide variety of carcinogens to cellular proteins. Examples of nucleophilic centers present in protein which may be attacked by activated carcinogens are the sul-

Table 1
BROAD CLASSES OF CHEMICAL COMPOUNDS KNOWN TO
PRODUCE CANCER

Alkylating agents	Mustards, epoxides, ethyleneimines, strained-ring lactones, alkanesulfonates, nitrosamides
Aromatic amines	Nitrocompounds, heterocyclic derivatives, beta-naphthylamine, N-acetylaminofluorene
Azo compounds	Aminoazo compounds
Chlorohydrocarbons	Carbon tetrachloride
Hormones	Estrogens, especially diethylstilboestrol and estrone
Hydrazides and hydrazines	Dimethylhydrazine
Metals and other inorganics	Asbestos, nickel, etc.
Mycotoxins (multi-ring compounds)	Aflatoxins
N-Nitroso compounds	Aliphatic rather than aromatic
Polycyclic aromatic hydrocarbons	Ring N-, O-, and S- substituted derivatives
Radiochemicals	Bone-seeking isotopes
Silica compounds	Asbestos
Triazides	—

Information modified in part from Clayson, D. B., *Carcinogenesis Testing of Chemicals*, Goldberg, L., Ed., CRC Press, Boca Raton, Fla., 1976, 80. With permission.

fur of methionine and cysteine, the N1 and N3 atoms of histidine, and the C3, possibly O^4 atoms of tyrosine. Tryptophan is another possible but uncharacterized site of attack. Another important factor is that the transfer of bound metabolite from one center to another is a distinct possibility. Very little work has been done on the chemical nature of naturally occurring cellular carcinogen:protein complexes. The most clearly characterized example is the formation of methionine derivatives after azo dye and acetylaminofluorene (AAF) administration.[8]

Thus, it has been known for some time that many classes of chemical carcinogens become bound to DNA, RNA, and proteins in the cells of target tissues.[9-11] However, despite much intensive research, the critical macromolecular target has not been unequivocally identified. Most recent studies have been heavily concentrated on demonstrating interactions with DNA.[12] Aberrations in nuclear morphology and function found during carcinogenesis, together with bizzare biochemical changes and loss of growth control, have been attributed to modification or damage of the DNA template by carcinogens. This has led to a plethora of predictive assays for carcinogenic potential, such as those based on mutagenesis.[13] Although it is highly likely that certain types of DNA damage can give rise to cancer (e.g., the case of *Xeroderma pigmentosum*[14]), it is equally likely that the nuclear proteins could play a major role in malignant transformation. Alterations in DNA:protein complexes is a plausible carcinogenesis mechanism for the following reasons:

1. Every cell in the body is thought to be totipotent[15] and therefore contains the same informational DNA content. It follows that differentiation a form of mutation, must proceed without loss of DNA templates.
2. The DNA is in initimate association with, and is well protected by, proteins *in situ*. The protein to DNA ratio in isolated chromatin is usually of the order of 2:1[16] and in most nuclei is probably in excess of 4:1.
3. All known carcinogens that react with DNA also react with proteins. In many studies (quoted later), the binding of chemical carcinogens to nuclear proteins in vivo has been shown to be much greater than to the DNA. In other cases the binding to DNA in vivo is relatively small, other nuclear macromolecules being more extensively attacked.

Table 2
EXAMPLES OF CARCINOGEN REGIMENS COMMONLY USED FOR BIOCHEMICAL INVESTIGATIONS

Carcinogen	Animal	Typical method of administration	Low level dose	Target organ and incidence	Latent period (weeks) before appearance of tumor
Nitrosamines					
Diethyl or dimethyl	Rat M or F	Oral	40—50 ppm (in water)	Liver 100%	20—35
Aromatic amines					
N-2-Fluorenylacetamide (AAF)	Rat M or F	Oral	80 ppm (in diet)	Liver 30%, breast 50%	60—90
	Mouse M or F	Oral	240 ppm	Liver	90
Azo compounds					
3'-methyl-4-dimethyl amino-azobenzene (3'-MeDAB)	Rat M	Oral	0.06% (in feed)	Liver 55%	37
Polycyclic aromatic hydrocarbons					
7,12 Dimethylbenze[a]anthracene	Rat F	Oral	15—20 mg	Breast 100%	12—16
	Mouse M or F	Skin	75 mg	Skin	10—25
Mustards					
Nitrogen mustard	Mouse M or F	IP	0.21 mg/kg	Lung 40%	39
Uracil mustard	Rat F	IP	11.5 mg/kg	Breast 55%, lung 10%, lymphoma 10%, peritoneum 10%	71
Thioacetamide	Mouse M or F	IP	8 mg/kg	Lung 100%	24
	Rat M	Oral	0.25%	Liver 85%	12
Urethane	Mouse M or F	IP	10 mg	Lung 100%	24
3-Aminotriazole	Mouse M or F	Oral	300—2000 ppm	Thyroid, liver 45—65%	78

Modified from Weisburger, J. H., *Carcinogenesis Testing of Chemicals*, Goldberg, L., Ed., CRC Press, Boca Raton, Fla., 1976, 31. With permission.

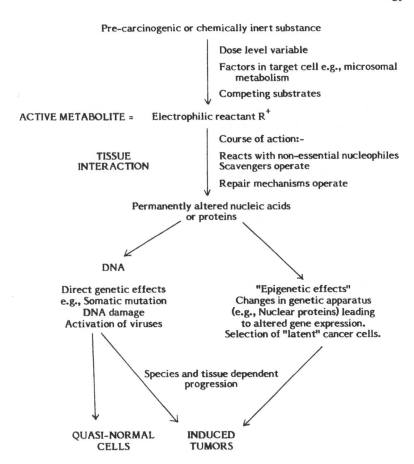

FIGURE 1. Scheme showing currently accepted postulated mechanisms of chemical carcinogenesis.

4. Nuclear protein-carcinogen adducts are seemingly stable and long-lived.

5. A large fraction of the DNA consists of repeated sequences which are neither transcribed nor translated. When a carcinogen has been shown to be bound to DNA this may mean nothing, since the bulk of the template is permanently shut off. Binding to active gene areas needs to be demonstrated.

6. The presence, stability, and persistence of modified bases does not, in many cases, seem to correlate with carcinogenic activity.[17-19]

7. Modified or damaged DNA bases can be, and are readily removed by repair; although it is recognized that incorrect copying can occur here.

8. No one has clearly demonstrated the loss of a specific DNA sequence or a DNA sequence coding for cell specific products, during, or relevant to carcinogenesis. If DNA damage is responsible for neoplasia one would not expect the resultant tumor to be totipotent. It is not known if this is generally true, but one form of tumor cell line, namely teratomas, definitely are totipotent (also some types of virally induced tumors).

9. Some work[20,21] suggests that the target size for neoplastic transformation appears to be considerably larger (20- to 25-fold) than that for mutation; indicating that protein to DNA interaction (such as those involving allosteric changes) could be important.

Thus, there is a strong possibility that the so-called "epigenetic effects" leading to altered patterns of transcription and modified phenotype could represent a major mechanism in malignant transformation and leads one to the conclusion that there is every justification for greater intensity of effort in examining the interactions and effects of carcinogens with nuclear proteins. Biochemical work has been carried out along these lines for the last 30 years, but much of it is fragmentary, inconclusive, and subject to the technical problems; particularly those involved in obtaining meaningful protein preparations from the nuclei and isolated chromatin of target tissues. By far the bulk of the work has been done on liver (particularly in rats and mice) presumably because techniques for the isolation of nuclei and components were the best advanced (at the time) and sufficient material could be obtained for analysis.

In the following sections an attempt has been made to summarize the relevant work on in vivo investigations on nuclear proteins during carcinogenesis. Numerous papers on tissue culture, model, or reconstituted systems, have not been included since their relevance to a dynamic multistep process, such as carcinogenesis, is not clear. There is also a great deal of published data on the effects of chemical carcinogens on the nuclear basic proteins — the histones. The question of carcinogen-induced changes in histone content, turnover structure, micromodification, and metabolism have been dealt with in detail in recent reviews, especially with regard to current ideas of nucleosome structure.[22,23,29] The importance of such events in carcinogenesis has yet to be ascertained. These depend on several factors, such as the role, if any, of nucleosomes, in the transcriptionally active areas of the genome (see Kornberg[24] for discussion). One of the current hypotheses is that specific nonhistone proteins (NHP) control the distribution of nucleosomes in a nonrandom fashion.

It is evident from much recent work,[29] and that described in other parts of this book, that the nonhistone proteins play a vital role in the control of gene expression. With the introduction of new improved methods of nuclear isolation, NHP extraction, and two-dimensional analysis, one hopes that the findings of earlier work reported here will be improved and consolidated in the near future.

II. WORK ON NONHISTONE PROTEINS USING ANIMAL CARCINOGENESIS EXPERIMENTAL SYSTEMS

Unless otherwise stated, the following data refer to liver carcinogenesis work. No attempt has been made to compare the various techniques employed to investigate NHP changes and properties; it is probable that each method has its individual merits; hopefully in the near future some attempt will be made to standardize the methodology employed so that results can be more meaningfully compared.

A. *N*-(2)-Acetylaminofluorene

An amine that has been extensively used for liver carcinogenesis studies is 2-acetyl-aminofluorene (2-AAF). Its active carcinogenic species formed in vivo, the "proximate carcinogen", has been shown to be *N*-hydroxy AAF. Esters of this compound provide the "ultimate carcinogen", one of the most reactive of which is the sulfate. This is formed in the liver as shown in Figure 2. The sulfate and other esters, react with critical nucleophiles in the cell. These reactions can occur nonenzymatically at pH 7.[8] In proteins, tyrosine, tryptophan, cysteine, and methionine residues have been shown to be attacked to give covalent adducts and this can cause peptide bond cleavage.[25] A hydrolysis product, 3-methyl-mercapto-AAF has been obtained from the proteins of the liver of AAF-treated animals, presumably formed from a methionine adduct, as in the case of the azo dyes.

Binding to nucleic acids has also received much attention, particularly to the DNA, where the formation of stable adducts such as *N*-guanin-8-yl-2-AAF and 3-(guanin-*N*²-yl)-2-AAF have been shown to result in marked conformational changes in the genetic material.[26]

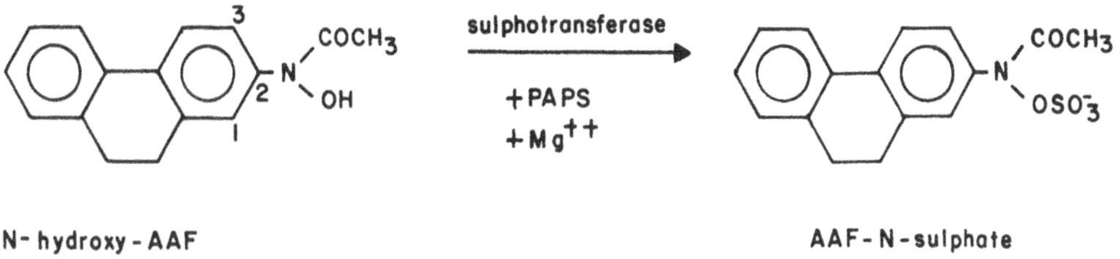

N-hydroxy-AAF   AAF-N-sulphate

FIGURE 2.   Formation of an "ultimate" carcinogen from *N*-hydroxy-AAF.

In studies on the mode of action of AAF the tendency of many workers has been to use a "shortcut", that is, to administer the "active" form of AAF, the *N*-hydroxy derivative. Most of the nuclear studies performed have been concerned with the binding characteristics of this carcinogen. The results of these investigations are summarized in Table 3.

It can be seen that there is considerable variation in the findings of binding studies. Attempts at chromatin fractionation have also met with limited success. In this connection, a number of in vitro studies of the binding of the active forms of AAF to isolated chromatin have appeared in the literature, but the interpretation of these results in terms of carcinogenic mechanisms must be treated with caution.

B. Aflatoxins

The aflatoxins are a group of substituted coumarins containing a fused dihydrofurofuran moiety and are produced as secondary fungal metabolites by specific strains of *Aspergillus flavus* and *A. parasiticus*. They are found as contaminants in food products causing concern since they are potent hepatotoxic and hepatocarcinogenic agents (especially Aflatoxin B_1) in a number of animal species.[41]

Microsomal metabolism of these compounds leads to the formation of metabolites, possibly via epoxides or diol-epoxides, which form covalently linked adducts with nucleophilic sites in target cell nucleic acids and proteins in vivo. Aflatoxin-activating enzymes have also been reported in the nucleus (rat liver[42]). The morphological macrosegregation and disruption of the nucleolus is the most noticeable effect after the administration of this carcinogen. A concomitant inhibition of RNA synthesis is also well documented.[43,44] Some evidence indicates that Aflatoxin B_1, in particular, may affect RNA synthesis by directly influencing nuclear polymerase activities.

In a recent paper, Groopman et al.[45] reported on the binding of ^{3}H Aflatoxin B_1 to rat liver nuclear fractions (single IP dose). Although the bulk of radioactivity quickly became associated with the DNA (approximately 80%) some was bound to nuclear proteins. The apparent removal of radioactivity from the DNA was twice as fast as that of the chromatin proteins. Histone H1 was established as a major target; however, the bound radioactive aflatoxin turned 3 to 4 times faster than this histone fraction. An unidentified labeled fraction (DNA-protein adduct?) was also shown to be present by SDS electrophoresis.

Although the evidence to date seems to indicate that the carcinogenic action of aflatoxins is mediated via DNA modification, the importance of nuclear protein interactions cannot be ruled out at this stage.

C. Azo Dyes

The azo dyes represent a well-characterized group of chemical carcinogens acting on the liver (hepatocarcinogens), the chemistry and biochemistry of which have been extensively researched by Japanese and American workers (for reviews see References 46-49). The chemicals most commonly used for experimental studies are the derivatives

Table 3
NUCLEAR METABOLISM OF N-ACETYLAMINOFLUORENE (AAF)

Alkylation and Binding

Compound used	Dose, conditions, duration, etc.	Result	Ref.
[9-^{14}C]-AAF	Seven injections of labeled carcinogen	Stable protein adducts found; studies mainly histone binding (H3 & H4 max)	27
[9-^{14}C]-N-OH-AAF	Two injections (0 & 24 hr) killed after 48 hr	Four times more bound to NHP compared to histones (based on crude fractionation)	28
[9-^{14}C]-NOH-AAF	One injection (up to 90 min duration)	12 pmol of carcinogen bound per milligram of NHP (after 60 min)	62
[9-^{14}C]-N AAF	Two doses i.p. killed after 48 hr	Approximately 80% of AAF bound to NHP; 43 pmol carcinogen per milligram; urea-soluble NHP; 55 pmol/mg residual NHP; persistent binding to NHP over 2 weeks; NHP analyzed by isoelectric focusing: complex pattern obtained; very little bound to DNA	30
^{3}H-N-OH-AAF	Intraperitoneal injection; liver perfused before nuclear isolation; binding studied for up to 11 days	Maximum binding on day 1 dropping rapidly; ploidy classes of liver nuclei isolated; tetraploid bound 2× more than diploid; binding to DNA much lower than to protein or RNA	31
^{3}H-N-OH-AAF	Single injection; examined up to 10 days	Binding liver chromatin fractions studied highest in RNA; initially more bound to "euchromatin" the "heterochromatin" but later (10 days) more found in the latter; balance could be altered by feeding AAF in diet casting doubt on meaning of single dose N-OH-AAF experiments; carcinogen-DNA adducts identified; tritium exchange occurs?	32
[9-^{14}C]-N-ACETOXY-AAF	In vitro with calf thymus chromatin	Findings suggested formation of a protein-carcinogen — DNA complex	33
AAF N-OH-AAF AF (2 aminofluorene)	In vitro and in vivo investigations following single injection	AF and N-OH-AAF more actively bound; DNA bound more per milligram than did protein	34

Template Recognition Transcriptional Effects, Polymerase Activity

Compound	Template Recognition	Transcriptional Effects, Polymerase Activity	Ref.
N-OH-AAF	Injection single i.p. dose; experiments on isolated liver nuclei	Rapid inhibition (80%) of Mg^{++} and Mn^{++} dependent RNA polymerase activity 15 min after carcinogen; template activities of chromatin and DNA not affected; females (not so susceptible to this carcinogen) gave negative results	35
N-OH-AAF		Similar results: inhibition of RNA polymerase activity represents an established example of a selective NHP target for a chemical carcinogen	36,37
N-OH-AAF		RNA pol II more sensitive than RNA pol I	38
N-OH-AAF	Single dose i.p. 2 hr duration before nuclei prepared	In vivo modification of RNA pol results in an alteracation of capacity to transcribe normal RNA product; product size transcribed from normal or carcinogen modified templates not affected	39
AAF	Fed in diet for up to 35 weeks	DNA polymerases examined α-form functions for de novo replication β for repair replication; waves of activity paralleling carcinogen activity, leading to tumor nodule formation	40

Note: References listed in order of publication; from earliest date to most recent.

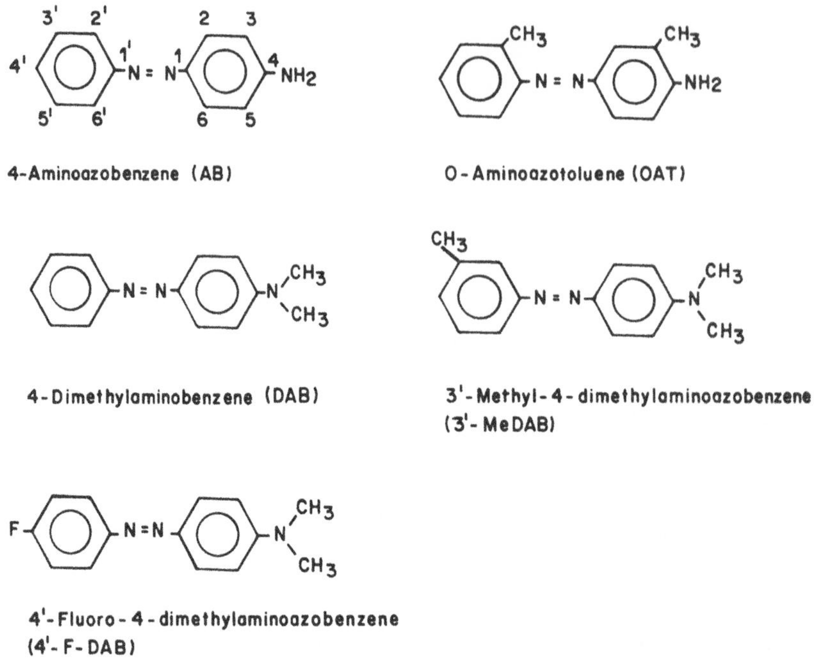

FIGURE 3.   Carcinogenic azo dyes and derivatives.

of 4-aminoazobenzene (AB) shown in Figure 3. The presence of at least one metabolically active *N*-methyl group substituted at the 4-amino group is necessary for carcinogenic potency.[49] Work on NHP change induced during azo dye carcinogenesis is summarized in Table 4.

D. Dimethylhydrazine (DMH)

In contrast to the bulk of carcinogenesis work, on nuclear metabolism (which is
based mainly on liver oncogenesis,) this carcinogen is used almost solely to examine
colon carcinogenesis, a disease particularly relevant to man. DMH is one of the most
potent and reliable colon carcinogens available for rodent experimental systems.

The colon is a difficult tissue to study since the colonic epithelium has a diversity of
cell types with complex cell kinetics. Furthermore, most of the cells have a relatively
short half-life. However, methods for cell separation and nuclear fractionation have
evolved which now permit biochemical investigation. Allfrey's group in New York has
been very active in this area (for reviews see Allfrey et al.[22,84]). The findings are as
follows:

Alkylation — DMH is almost certainly a carcinogen by virtue of its alkylating ability. Thus, Allfrey and Boffa[85,89] reported that alkylation of chromosomal proteins by
this carcinogen led to the formation of ε-*N*-monomethyl lysine, ε-*N*-dimethyllysine,
and N^G, N^G- dimethylarginine. The formation of methyl lysines in the NHP and N^G, N^G-
dimethylarginine in histones was regarded as aberrant.

Metabolism of nuclear proteins — A systematic study of nuclear nonhistone proteins
during the induction of colon adenocarcinomas by 1,2-dimethylhydrazine showed the
progressive accumulation of two nuclear protein classes of mol wt 44,000 (called frac-
tion-TNP$_1$) and 62,000 (TNP$_2$). The synthesis of these two protein classes was found
to be selectively accelerated within 4 weeks after administration of the carcinogen, long
before the appearance of pathological indications of malignancy.[86,87] Fractionation of
the tumor nuclei showed that proteins of the TNP$_1$ class were largely localized in the

Table 4
NUCLEAR METABOLISM OF AZO DYES

Alkylation and Binding

Compound used	Dose, conditions, duration, etc.	Result	Ref.
DAB	—	Covalent binding of dye to liver proteins	50
3'-MdDAB	Single large doses given	Azo dye binding to rat liver nuclei reported	51
3'-MeDAB	18—24 days, 0.06% w/w in diet	Nuclei isolated; contained bound dye	52
3'-MeDAB		Methionine adduct isolated from liver proteins after azo dye feeding	53
^{3}H-DAB	10 days, 0.06% diet	Histones, found to bind 66, and NHP 205 pmol of dye per milligram protein	54,55
^{3}H-DAB (ring A labeled)	Single injection 150 mg/kg	pmol of carcinogen per milligram protein found: (a) cytoplasmic 540, (b) nuclear protein 390, (c) DNA 14; maximum binding at 17—42 hr	56
3'-MeDAB } 2-MeDAB }	0.06% in diet 2 weeks	NHP bound 1½-3 times more dye in the case of the carcinogen compared with the noncarcinogen	57
3'-MeDAB	20 days on 0.06% (in diet)	3% of bound azo dye in nuclei; bulk extracted with isotonic saline, 50% dye associated with small class of near neutral proteins (by zonal electrophoresis)	58
^{3}H-DAB	Single dose i.p. 45 mg/kg	^{3}H-dye bound to nuclear macromolecules	59
^{2}H-DAB } ^{3}H-2-MeDAB }	Single dose i.p. 45 mg/kg	Higher binding to crude nuclear fraction in the case of the carcinogenic dye, 125 pmol bound per milligram NHP after 24 hr	60,61
^{14}C-DAB	Single injection, time course several hours	Maximum covalent binding at 60 min — 30 pmol/mg NHP — disappeared after 30 days	62
DAB } 2-MeDAB }	1—4 weeks dye feeding (0.06% in diet)	100 pmol of dye per milligram of nuclear protein found but no significant difference between 2 dyes	63

Table 4 (continued)
NUCLEAR METABOLISM OF AZO DYES

Alkylation and Binding

Compound used	Dose, conditions, duration, etc.	Result	Ref.
[3]H-3'-MeDAB	Single dose, 4 mg by stomach tube	Highest specific activity in an NHP fraction containing RNP particles, these correspond to informofers (mRNA) and contained most bound carcinogen	64
[14]C-DAB AB	Single i.p. dose (50 mg, approx 2.5 μmol)	Nuclei and chromatin binding studied; specific binding of carcinogenic dye (DAB) to NHP found (not DNA); very little binding of AB to nuclear macromolecules	65

Template Recognition, Transcriptional Effects, Polymerase Activity

Compound used	Dose, conditions, duration, etc.	Result	Ref.
DAB 3'-MeDAB	0.06% in diet	Up to 50% inhibition of nuclear DNA dependent RNA polymerases after carcinogen feeding	66,67
3'-MeDAB	0.06% in diet	Found increase in tightly bound nuclear DNA pols; 6—8 polymerase (similar to reverse transcriptase in properties) studied; activity doubled at 28—40 weeks but decreased gradually thereafter	68,69
(a) 3'-MeDAB (b) 3'-MeDAB and Chloramphenicol (CAP)	(a) 0.06% in diet, (b) 0.06% plus 12% CAP in diet	RNA pols I and II isolated from nuclei; significant increase in specific activities and/or amount of both enzymes on both regimens after 10 days; greatest when CAP present; total nuclear protein content also increased	70

Metabolism of Nuclear Proteins

Compound used	Dose, conditions, duration, etc.	Result	Ref.
DAB	22 weeks feeding (0.06% in diet)	In vitro incorporation of labeled precursors into nuclear macromolecules studied	71
DAB	Continuous feeding (0.06% in diet)	50% decrease in aspartic acid content of DNA-associated proteins observed	72

Compound	Treatment	Effect	Ref.
3'-MeDAB	Intraperitoneal injection	100% increase in synthesis of total nuclear proteins reported	73
3'-MeDAB 2-MeDAB	36 weeks feeding (0.06% in diet)	Chromatin prepared from isolated nuclei; significant decreases in NHP content of liver of carcinogen treated animals (no change 2MeDAB)	74
3'-MeDAB	Single dose 250 mg/kg body weight (killed at 40 hr)	Chromatin fractions prepared; differences found between carcinogen and noncarcinogen preparations	75
DAB 2-MeDAB	Continuous feeding (0.06% in diet)	Complex effects on turnover of nuclear proteins found	76
3'-MeDAB	Continuous feeding (0.06% in diet)	Stimulation of incorporation of ^{14}C amino acids into all nuclear protein fractions except RNP which was reduced; little change in NHP electrophoretic patterns	77
3'-MeDAB	Continuous feeding (0.06% in diet)	Absence of 1 of major 30 RNP proteins (mol wt 42,000) after 10 weeks of feeding (not observed when noncarcinogenic dye used)	78,79

Protein Micromodification — Phosphorylation

Compound	Treatment	Effect	Ref.
3'-MeDAB	0.06% in diet up to 106 days	Intensive phosphorylation of several NHP in chromatin during feeding of carcinogen; both cytoplasmic and chromatin bound phosphoprotein kinase activities increased several-fold	80
3'-MeDAB	0.06% in diet up to 67 days	Increase in rate of phosphorylation of NHP observed after 26 days	81
3'-MeDAB	0.06% in diet	Chromatin NHP phosphorylation patterns found to vary in control, regenerating liver and carcinogen-treated animals	82

Immunological Changes

Compound	Treatment	Effect	Ref.
3'-MeDAB	0.06% in diet	Azo dye feeding found to alter the immunological specificity of chromosomal NHP-DNA complexes (NHP binding protein responsible?)	83

Note: References listed in order of publication; from earliest date to most recent.

dividing cell population of the tumor, while TNP_2 was enriched in the nondividing cells.[88] Neither protein was present in the colonic epithelial nuclei of normal adult rats, and they were not detectable in the fetal colon. Similar proteins were found to be present in the nuclei prepared from human colonic adenocarcinomas, but not in nuclei from nonmalignant intestinal polyps.[85] The tumor-specific protein class TNP_1, shows high affinity for DNA, while TNP_2 does not bind to DNA. Limited digestion of tumor chromatin with DNase I resulted in a preferential release of TNP_1 and no loss of TNP_2.[85] The results strongly suggest an association of TNP_1 with extended, template-active regions of the tumor nucleus. The heterogeneity of the 44,000 mol wt protein class was investigated by two-dimensional gel electrophoresis. Isoelectric points for TNP_1 were found to range from pH 4.85 to 5.25, in agreement with the predominance of acidic amino acids over basic amino acids in its composition.[84,86] Later, both classes were found to be complex when analyzed by two-dimensional electrophoretic techniques.[90] These authors state that "These latest results confirm the view that alterations in nuclear protein complement distinguished DMH-induced adenocarcinomas of the colon and suggest the presence or absence of particular nuclear protein classes may signal changes leading to malignant transformation."

Immunological changes — Recent work by Chiu et al.[91,92] has shown that administration of DMH to rats produced an early change in the immunospecificity of colon epithelial chromatin. Several lines of experimental evidence indicated that the nuclear antigen was not a carcinoembryonic-antigen-like substance. Common antigens were also present in human adenocarcinomas.

E. Ethionine

The hepatocarcinogen ethionine, $HOOC \cdot CH \cdot (NH_2) \cdot CH_2CH_2S \cdot C_2H_5$, as an analogue of methionine, is thought to act by ethylation of vital cellular macromolecules. In fact it is difficult to envisage any other mode of action. Ethylation of mainly RNA and protein has been demonstrated but not the DNA of the target cell. The biochemistry of ethionine carcinogenesis has been reviewed by Farber[93] and Stekol.[94] This compound is a potent inhibitor of protein synthesis, causes a depression in cellular ATP levels, and inhibits RNA formation in vivo; the relationship of these events is not clear. Smuckler and Koplitz[95] found inhibition of RNA synthesis in isolated nuclei from ethionine-treated rat's liver. There were no differences in nuclear composition, nuclear RNase, or in the capacity of chromatin or DNA to act as a template for exogeneous RNA polymerase. Their results suggested that ethionine treatment had altered the RNA polymerase enzyme or the enzyme-chromatin interaction. However, Farber et al.[96] found that although ethionine caused inhibition of RNA synthesis and fragmentation of nucleoli, this could be reversed by adenine.

Micromodification of nuclear proteins by ethylation might be expected to change their interaction with nucleic acids. However, methylation certainly occurs naturally. Ethylation of liver nuclear proteins has been shown to occur when (Et-^{14}C) ethionine was injected into rats.[97] Later work by Friedman et al.[98] showed highly selective ethylation of the nuclear proteins soluble in 0.14 M NaCl. On hydrolysis of these proteins an ethyl analogue of methylguanidino arginine was tentatively identified. It would be interesting to know if this protein material was associated with the RNP particles sometimes extracted from nuclei with diluted saline. Proteins associated with RNP particles (which contain mRNA) have been shown to be methylated. Orenstein and Marsh[99] examined the ethylation and methylation of rat liver chromatin associated proteins after a single dose of labeled L-ethionine or L-methionine. By using ^{35}S-labeled compounds and those labeled with ^{14}C or ^{3}H in the alkyl group (ethyl or methyl) they hoped to distinguish between alkylation and metabolic incorporation into the nuclear proteins. They concluded that methionine was incorporated 16 to 26 times more readily

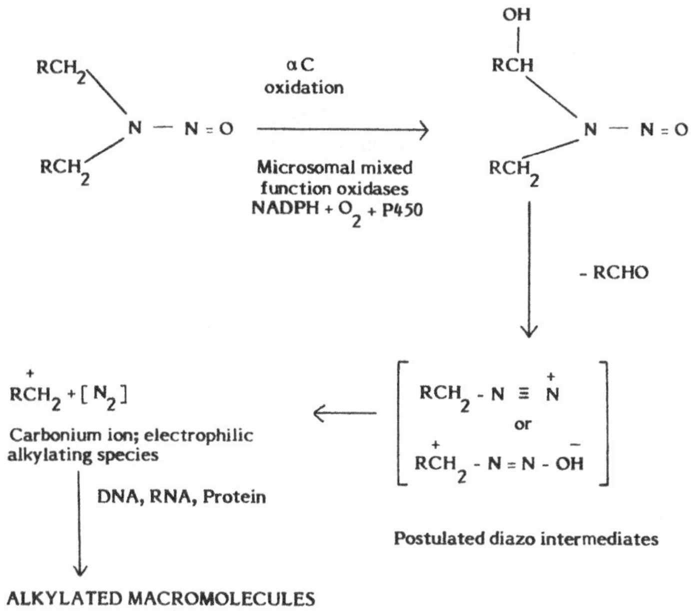

$$RCH_2 \diagdown$$
$$N - N = O \xrightarrow[\substack{\text{Microsomal mixed} \\ \text{function oxidases} \\ \text{NADPH} + O_2 + \text{P450}}]{\substack{\alpha C \\ \text{oxidation}}}$$
$$RCH_2 \diagup$$

$$\substack{OH \\ | \\ RCH} \diagdown$$
$$N - N = O$$
$$RCH_2 \diagup$$

$$- RCHO$$

$$\overset{+}{RCH_2} + [\overset{+}{N_2}]$$

Carbonium ion; electrophilic
alkylating species

$$\left[\begin{array}{c} RCH_2 - N \equiv \overset{+}{N} \\ \text{or} \\ \overset{+}{RCH_2} - N = N - \overset{-}{OH} \end{array} \right]$$

| DNA, RNA, Protein

Postulated diazo intermediates

ALKYLATED MACROMOLECULES

FIGURE 4. Postulated breakdown mechanism and biological mode of action of nitrosamines.

than ethionine into all the nuclear proteins studied. Lysine-rich histones incorporated more methionine and ethionine than did arginine-rich histones. Chromatin NHP accepted few alkyl groups but the methylation of residual NHP was considerable. As usual, NHP showed a high rate of overall synthesis compared to the histones.

The methylation and ethylation of nuclear proteins is obviously a complex story and could represent an important nuclear metabolic regulatory mechanism. It is of interest that recent data[89] indicated that specific nuclear proteins associated with Hn RNA contain high amounts of N^G, N^G-dimethyl arginine. These proteins could represent selective targets for this type of carcinogen.

F. Nitrosamines and Nitrosamides

These chemicals represent a group of very powerful biological alkylating agents which are potent carcinogens and mutagens at low levels of exposure. They may represent an important group of environmental carcinogens.[100] Nitrosamines do not give rise to alkylating species until metabolized by selected cells. By contrast, the nitrosamides are generally reactive chemicals which have a short half-life in aqueous solutions (especially at alkaline pH) and in vivo. Although the potency of these groups of chemicals as a class of carcinogens is beyond dispute, the exact mechanism of action is still unknown. It has been postulated for some time that the active alkylating species are produced from nitrosamines after enzymatic oxidative dealkylation to give a diazo intermediate. The latter breaks down to an alkyl or aryl carbonium ion (Figure 4). This can then react with active nucleophilic centers in the cell. In the case of compounds with long side chains, oxidation on carbon atoms other than the α, may occur, and, in the case of cyclic nitrosamines the ring may be opened. It is thought that nitrosamides do not require enzymatic activation to produce active carbonium ions. Systems of nitrosamine carcinogenesis involving high to low dose regimens have been extensively described by Druckrey's group.[101] Unfortunately, many investigators of the nuclear aspects of hepatic carcinogenesis (the most commonly used system) induced by nitrosamines have relied on single, or multiple high doses capable of inducing extensive necrosis in addition to the promotion of neoplastic change.

Dimethylnitrosamine (DMN) and diethylnitrosamines (DEN) have been the most popular nitrosamines used for biochemical investigations. Damage to the nuclei of the cells by these compounds is well documented. Like many other hepatoxic substances, the nitrosamines cause an increase in nuclear size and also increased polyploidy.[102] Chromosomal damage and abnormalities have been reported during carcinogenic administration of nitrosamines.[103] Evidence for change in the physical properties of chromatin (modification of melting curves) after DMN and DEN treatment has been given by Stewart and Farber.[104] Nicolini et al.[105] showed that distinct alterations in the conformation of rat liver chromatin (increased ellipticity in circular dichroism spectra and increased ability to bind ethidium bromide) occurred after a single dose of DMN (5 mg/kg). The maximum change was induced after 3 days and was not repaired for 14 days.

Although various cellular macromolecules undergo alkylation after administration of N-nitroso-compounds, for some 16 years principle consideration has been given to alkylation of the nucleic acids as the key event relating mutagenesis to carcinogenesis (see Magee[106] and Montesano and Bartsch[107] for discussions). However, alkylation of protein molecules does occur[108] such as on the 1 and 3 groups of histidine, the -SH of cysteine, and sulfur of methionine, etc. To date the alkylation of proteins by the N-nitroso-compounds, in particular the nuclear proteins, has received little attention compared to the nucleic acids. A summary of NHP studies directly relevant to nitrosamine carcinogenesis is given in Table 5.

G. Polycyclic Aromatic Hydrocarbons (PAH)

Some of the first chemical carcinogens to be identified belong to this group, viz dibenz [a,h] anthracene and benzo [a] pyrene (BP); they are classically skin carcinogens. It was soon recognized that not all compounds with a similar structure were carcinogenic.[3,127] Many of the carcinogenic species were found to be products of fossil fuel combustion and are therefore classified as important environmental pollutants. Over the last two decades various workers have demonstrated the covalent binding of different carcinogenic PAH to DNA[10,12,128,129] and as a result, it has often become axiomatic to correlate these observations with the critical event in carcinogenesis (see Reference 130 for short review).

Binding of PAH to protein has also been shown, but whether the binding that occurred was predominantly to protein or nucleic acid seemed to be a funcion of the individual PAH. However, following up either work[131] various groups have investigated the binding of PAH to nuclear macromolecules. This research is summarized in Table 6.

Obviously these rather inert chemicals do bind to nuclear proteins in a noncovalent fashion and could therefore act as gene activators in a similar fashion to the steroids (and related compounds). However, this effect would not be expected to produce the permanent type of change necessary to induce neoplasia. It is generally agreed that the covalent carcinogenic reaction of PAH with critical macromolecular targets proceeds through the formation of short-lived arene oxide intermediates such as diol epoxides.[129,142] These are believed to be the proximal or ultimate carcinogens and examples of probable compounds to be found in vivo are given in Figure 5. These are produced from PAH by the cytochrome P-450 mediated aryl hydrocarbon hydrolase system present in the cell;[10] phenols, transdihydrodiols, quinones, and water-soluble glutathionine conjugates are formed, some of which are potent mutagens to bacteria. This enzyme system is present not only in the microsomes but also within the nucleus.[143] It is induced by prior administration of a number of substances such as 3-methyl-cholanthrene.[133]

Table 5
NUCLEAR METABOLISM OF NITROSAMINES AND NITROSAMIDES

Alkylation and Binding

Compound used	Dose, conditions, duration, etc.	Result	Ref.
^{14}C-DMN	Single dose 30 mg/kg	Methylation of liver and kidney nuclear proteins S methylcysteine, 1-methyl histidine, 3-methyl-histidine, and ε-N-methyl lysine formed; no correlation found between extent of alkylation and carcinogenic effect	109
^{14}C-DMN/DMN	Single doses (2 mg/kg body weight); ^{14}C-formate also used	Pattern found to be complex, normal and abnormal methylation sites examined; demethylation at both sites within NHP but lower in "loosely bound" NHP after DMN; latter fraction also exhibited greatly increased turnover after carcinogen given	110
1-Methyl-1-nitro-sourea and 1-Methyl-3-nitro-1-nitrosoguanidine	Tissue culture of L1210 cells; incubation 60 min; 0.038 to 3.8 M concentrations of carcinogen	Many NHP and all histones alkylated (modified); acid soluble NHP 2× greater than histones	111
N[methyl-^{14}C] N'-nitro-N-nitrosoguanidine (^{14}C-MNNG)	AKR mouse embryo cells in culture	Selective binding to NHP shown in SDS profile (similar to PAH- see Table 6)	112

Template Recognition, Transcriptional Effects, Polymerase Activity

Compound used	Dose, conditions, duration, etc.	Result	Ref.
DMN MNU	5 mg/kg (DMN) 80 mg/kg (MNU)	Physicochemical alterations found in the conformation of rat liver chromatin; maximum change after 3 days, not repaired by 14 days	105
DMN	Single dose 40 mg/kg	50% decrease in RNA synthesis due to inhibition of polymerase activity	113
DMN	One dose after partial hepatectomy	Inhibition of induction of DNA polymerase synthesis	114
DMN DEN	10 mg/kg (DMN) 100—200 mg/kg (DEN)	Alterations in thermal stability of chromatin observed	115

Table 5 (continued)
NUCLEAR METABOLISM OF NITROSAMINES AND NITROSAMIDES

Template Recognition, Transcriptional Effects, Polymerase Activity

Compound used	Dose, conditions, duration, etc.	Result	Ref.
DEN	Continuous feeding	Initial increase in *de novo* replication of DNA (increase pol α); gradual increase in pol β (repair) correlating with carcinogen feeding	116
NMU		DNA pols I and III inhibited	117
Metabolism of Nuclear Proteins			
DEN	Single dose 280 mg/kg body weight	No change in rate of NHP synthesis for up to 32 days despite very marked stimulation of histone synthesis at 4 to 8 days	118
DEN	3 mg/kg/day in drinking water	Isoelectric focusing and SDS electrophoresis of NHP showed changes in distribution of NHP thiol components	119
DEN	3 mg/kg/day in drinking water	Chromatin fractions prepared; after 70 days a polypeptide of 17,500 mol wt found to be missing in a euchromatin fraction	120
DEN	Single injection 20 mg/kg	Increased synthesis of NHP for up to 3 hr; no differences in NHP composition could be found	121
N-ethylnitrosourea	75 mg/kg = carcinogenic dose for brain tumors; examined 5 days afterwards	Increased rate of chromosomal protein synthesis and changes in DNA affinity for NHP; differences in electrophoretic profiles described (brain tissue)	122
N-nitrosomorpholine	10 mg/kg/day in drinking water	Chromatin prepared — trouble with proteases; very little change in NHP found until hepatomas appeared	123
DMN	Single dose 9 mg/kg body weight given i.p., 6 hr after partial hepatectomy (carcinogenic)	HMG-type NHP examined; considerable reduction in incorporation of ^{3}H-lysine into these proteins	124

| DEN | 40 mg/l in drinking water | No qualitative changes in pattern of chromatin-associated NHP; quantitative changes 4th to 10th week; increases 41—47 and 51—64 kdalton fractions; decreases 27—32 and 47—51 kdalton fractions | 125 |

Protein Micromodification — Phosphorylation

| DEN | Continuous feed 10 mg/kg/day | Preneoplastic stages characterized by oscillating phosphorylation and dephosphorylation patterns of nuclear proteins | 126 |

Note: References listed in order of publication; from earliest date to most recent.

Table 6
NUCLEAR METABOLISM OF POLYCYCLIC AROMATIC HYDROCARBONS

Alkylation and Binding

Compound used	Dose, conditions, duration, etc.	Result	Ref.
³H-7,12 DMBA	Single i.p. injection (rat liver)	Covalent binding demonstrated to both histones and NHP; early fractionation procedure	62
³H-B (a)P	Incubated with isolated rat liver nuclei	Aralkylation of DNA, histones, and NHP demonstrated; 4× more bound to NHP as DNA	132,133
³H-B (a)P	Same as above	NHP found to bind twice as much carcinogen as DNA but RNase sensitive fraction 3-5× more; true covalent binding shown	134
Many labeled PAH	AKR mouse embryo cells in culture	Binding of ³H-PAH to various nuclear and chromatin fractions, a metabolically dependent process; showed covalent and noncovalent binding occurs; preliminary analytical experiments on aralkylated NHP indicated selective (covalent) binding species present (e.g., 65—75 kdaltons area)	135
³H-B (a)P	Model calf thymus system (nuclei) in vitro incubation NADPH and rat liver microsomes present	Carcinogen bound to nuclease accessible regions-nucleosome: spacer areas?	136,137
B (a)P-DIOL Epoxides	Model studies on nucleosome regions (chicken erythrocytes)	Binding to HMG-NHP of internucleosomal DNA (linker region) demonstrated	138,139
³H-B (a)P	Hamster embryo fibroblasts (culture)	Separation achieved of bound BP and derivatives to various nuclear classes-DNA, RNA, and protein; vast majority of carcinogen binding to protein rather than to nucleic acids; dihydrodiols bound mainly to histones (H2A and H3)	140,141

Note: References listed in order of publication; from earliest date to most recent.

Numerous studies have been made of the binding of PAH to the DNA, RNA, and proteins. Only a relevant portion of this work has been given in Table 6. Despite these investigations, almost nothing is known yet of the nature and binding site of the metabolites of PAH to the nuclear proteins.

It should be pointed out that a large proportion of the studies reported in Table 6, and others in the literature, are based on hypothetical or model systems. These include

1. Incubation of labeled carcinogen or "active species" with "whole" tissue culture cells — not necessarily "target" cells for transformation.
2. Incubation of labeled carcinogen or "active species" with isolated nuclei, chromatin, etc. This is difficult to carry out properly since "active species" are short-lived.
3. Nuclease digestions of carcinogen-bound "model" chromatins such as chicken erythrocyte or calf thymus.

Some of these investigations undoubtedly serve to confuse rather than to clarify the situation and therefore limited reference has been made to them in the table. It is evident from some of this work that the accessibility of DNA sequences to the PAH carcinogens (and others, e.g., AAF) correlates well with accessibility to nuclease probes

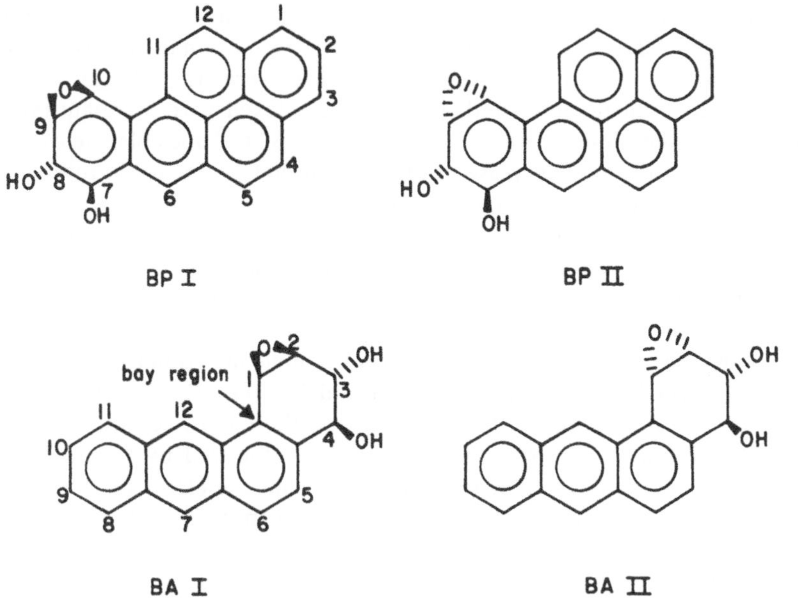

FIGURE 5. Possible diol epoxide of Benzo [a] pyrene (BP I and II) and Benzo [a] anthracene (BA I and II).

of chromatin structure. The shielding of DNA through associations with histones and NHP, such as the HMG group, could be a major factor in the carcinogenesis process and therefore merits further investigations.

H. Thioacetamide

Thioacetamide (CH_3CSNH_2) produces an increase in the size of the nucleoli of liver cells which is accompanied by an increase in nucleolar RNA and protein.[144] In addition, an eight- to ninefold increase in the ribonuclease activity of isolated nucleoli has been reported 48 hr after a single dose of thioacetamide.[145]

Barton and Anderson[146] found that, after 11 days of thioacetamide treatment the protein content of isolated rat liver nuclei had increased dramatically. The bulk of the increase occurred in a fraction extracted in 0.2 N HCl (crude histone) and in the residual NHP of the chromatin. Other easily extracted nuclear proteins had decreased in quantity. Their histone to DNA ratio was twice that usually found in eukaryote chromatin. However, in a later paper[147] they examined their histone fraction more carefully and the new data indicated that contaminating NHP could have been present, but not in any quantity. In another paper, Muramatsu and Busch[148] reported increased labeling of histones during thioacetamide treatment.

A complex study was carried out by Gonzales-Mujica and Mathias[149] on changes in nuclear proteins from different classes of nuclei (isolated by zonal centrifugation), obtained from liver of normal and thioacetamide-treated rats. NHP were examined by isoelectric focusing and SDS electrophoresis techniques. Different classes of nuclei gave different heterogeneous NHP patterns. This was especially noticeable when stromal and parenchymal nuclei were compared. Drastic alterations in NHP patterns were found after thioacetamide treatment but these were complex and not uniform. However, chromosomal NHP displayed specific changes which were different in adolescent and young adult rats. They also reported a general increase in the easily soluble 0.14 M NaCl extract and chromosomal NHP but a decrease in amino acid uptake into the residual and easily soluble NHP.

In later analytical studies on NHP changes in rat liver nuclear proteins after thioacetamide treatment, Busch's group obtained a better resolution using a two-dimensional electrophoresis system. Unfortunately, the changes were still difficult to quantitate. Acid extracts of both nuclei[150] and nucleoli[151] displayed various quantitative differences in a two-dimensional pattern. In the case of of the nucleolar acid-soluble proteins, they noted an overall increase in the ratio of NHP to histones following thioacetamide treatment. Most spots that increased in size and density had electrophoretic mobilities similar to the proteins found in preribosomal ribonucleoprotein particles.

Very little has been reported on the binding of thioacetamide or its metabolities to proteins, but a recent paper[152] describes the presence of a protein-bound form of thioacetamide in liver nucleoli.

Leonard and Jacob[153] examined the effects of a single injection (50 mg/kg) of thioacetamide on the rat liver RNA polymerase I and II. The effects were complex but a three- to fourfold stimulation of the activity of partially purified bound nucleolar polymerase IB was found. The size of the product synthesized by bound polymerase I was not significantly altered. The data suggested that the effect of thioacetamide on RNA polymerase I is manifested by an increase in the amount of enzyme protein rather than by activation of preexisting enzyme molecules.

Some current papers by Smuckler's group describe the effect of thioacetamide feeding on the nuclear envelope from rat liver; nucleoside triphosphate activity was increased nearly 300% compared to controls but there was no change in protein kinase activity.[154] On the other hand, little change in polypeptide composition could be observed, despite large increases in protein content.[155] These authors propose that the triggered expansion of the nuclear membrane may be an important first stage in the action of carcinogens.

I. Urethane

The mechanism of action of urethane-ethyl carbamate ($H_2N \cdot COOC_2H_5$), remains obscure although it is structurally one of the simplest carcinogens.[156] This compound can induce many types of tumors, but when given in a single dose to 8- to 9-day-old B6AF1 mice a high incidence of multiple hepatomas occurs. The system has the advantage of minimal toxicity; morphological damage does not occur. Gronow and Lewis[157] found changes in liver nuclear protein metabolism were induced in this system. The NHP showed a significant decrease in the uptake of labeled amino acids 8 to 18 hr after giving the carcinogen; a slight stimulation was evident at 24 hr. Histone synthesis was unaffected. Isoelectric focusing analysis of the NHP revealed quantitative and qualitative differences in the synthesis of components present in the mixture. The most noticeable change in the staining pattern was an increase in a protein/polypeptide having a pI of 7.35 and the appearance of new bands at pIs of 7.85 and 5.55 in the 18-hr treated livers. ^{3}H-tryptophan labeling indicated that this was due to an increased synthesis of these components. After 24 hr of urethane treatment there appeared to be an increased rate of synthesis of some of the major components of the mixture, particularly in the pI 5.65 region. Possibly alkylation of the side chains of the NHP could explain these differences, as it was notable that little change occurred in the SDS electrophoresis patterns which showed molecular weight differences. In fact, Prodi et al.[158] have reported that urethane, with a radioisotope in the ethyl group, labeled rat liver nuclear proteins, giving eight times the specific activity of the RNA.

III. DISCUSSION

There is now very strong evidence that, in carefully conducted carcinogenesis binding studies, one finds the bulk of bound carcinogen in the nucleus attached to the nuclear proteins, particularly the NHP, rather than to the DNA. How these protein-bound molecules affect the transcription of the DNA, is largely unknown (except in the case of AAF inhibition of polymerase activity, see Table 3).

It is notable that not one single carcinogen-nuclear protein adduct has been clearly characterized to date. The isolation of such adducts should be a priority item for future investigations.

The fact that alkylation occurs so readily, and that many of the most potent carcinogens are simple ethylating or methylating agents, *must* point to the importance of this event in neoplastic transformation. As methylation is a naturally occurring, dynamic, biochemical event, the problem is a complex one to deal with experimentally. As a result very few investigators have attempted to do so.

A theory that aberrant nucleic acid methylation is a primary event in cancer formation was put forward as early as 1964 by Srinivasan and Borek[159] (discussed later by Gantt[160]). Generally, we know from nitrosamine carcinogenesis work that the observed methylation or ethylation of DNA fails to correlate with carcinogenic change.[17-19] This makes aberrant alkylation of NHP an equally likely mechanism of altering the balance of genetic expression towards neoplasia, leading one to the inevitable conclusion that more work needs to be done on this important aspect of cancer research.

Methylation and ethylation are also described as protein micromodification. These micromodifications can become an important factor when one considers the effect of carcinogens on overall NHP analytical patterns, either single or two-dimensional. In our experience, if carcinogenesis experiments are carefully performed, one observes little qualitative change in either isoelectric focusing or SDS patterns right up to the appearance of tumors. One can produce different quantitative patterns by juggling with chromatin fractionation and extraction procedures, but the overall pattern remains relatively unchanged (as judged by currently available techniques).

For example, in Figure 6A we show some two-dimensional gel patterns obtained from rat liver NHP after feeding rats with 5 or 10 mg/kg/day of DEN in the drinking water for 40 to 80 days. After isolation of nuclei, the NHP ($\sim70\%$ of the total present) was extracted with 8 M urea 50 mM phosphate pH 7.6. These NHP, when analyzed by isoelectric focusing followed by SDS electrophoresis,[162] displayed a consistent pattern, only a few small areas marked were changed on carcinogen feeding. However, residual chromatin (after 8 M urea extraction of nuclei) or nuclesome-like fragments (isolated after endonuclease digestion) revealed quantitative losses in residual NHP (Figure 6B).

There is, however, a possibility that these analytical patterns are artifactual and that the so-called NHP are not real "species" at all but combinations of relatively low molecular weight polypeptides joined by an unusual and as yet unknown chemical link (nonpeptide). We have presented evidence to support this contention.[161,162] This tends to indicate very strongly that we need to know more about the chemistry of the NHP before further pursuing the search for classical repressors in eukaryote systems. Furthermore, we have found no convincing evidence that the classical prokaryote repressor type molecules exist in eukaryote nuclear protein fractions.

Finally, it is true to say that considerable advances have been made over the last few decades in the elucidation of the metabolism and mechanisms of reaction of carcinogens with cellular constituents, but we still do not understand the mode of action of even one carcinogen. Obviously no single theory or mechanism can explain the bewildering series of facts surrounding neoplastic initiation, promotion, and transformation; nor the great variety and number of chemical, physical, and viral carcinogenic agents known to man. Despite the very strong tendency in recent years to regard DNA as the only viable critical target in carcinogenesis, this remains unproven. Indeed, there is much evidence to the contrary clearly pointing to a vital role for the gene controlling proteins present in the NHP fraction. Although temporarily "out of fashion" one

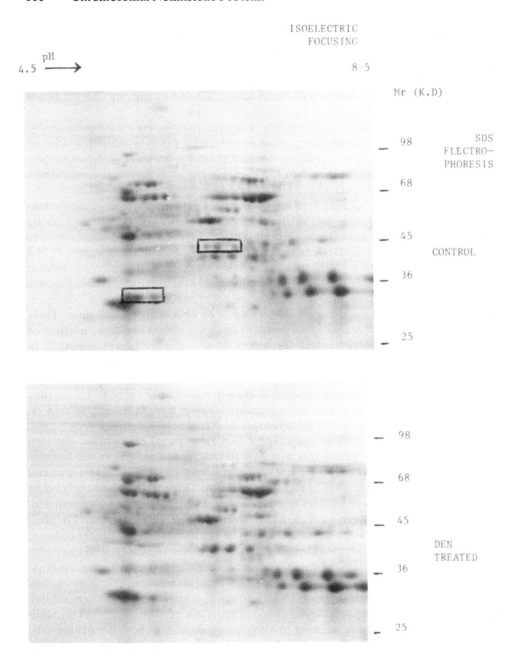

FIGURE 6A. Effect of DEN feeding on the NHP composition of rat liver nuclei. Two-dimensional gel separation of 8 *M* urea 50 *M* phosphate-soluble rat liver nuclear proteins (NHP).[162] The liver NHP from animals treated with either 5 or 10 mg/kg/day of DEN displayed the same patterns after 40 to 80 days. Note the loss of stained spots in the areas marked.

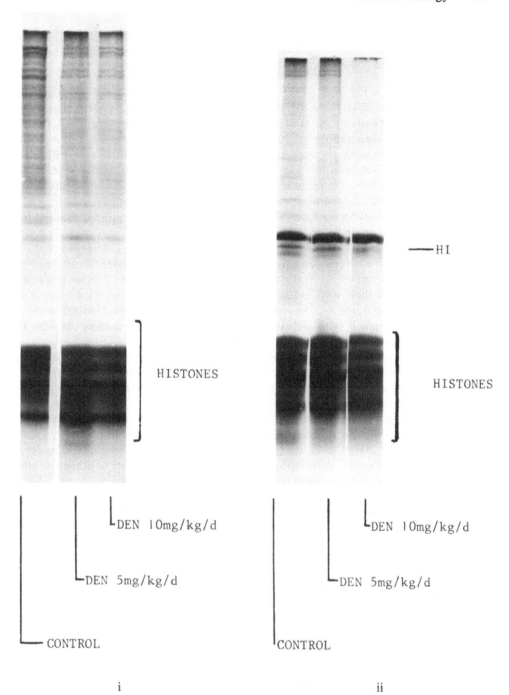

FIGURE 6B. Effect of DEN feeding on the NHP composition of rat liver nuclei. SDS electrophoresis of residual NHP (and histones) not extracted by buffered 8 *M* urea. (i) Nucleosome fragments isolated after endonuclease digestion and (ii) Residual chromatin (after 8 *M* urea extraction). All nuclear preparations were performed in the presence of the protease inhibitor PMSF. Note the diminution of stained bands in the NHP region (compared to the histones) after carcinogen treatment (40 to 80 days).

feels that the so-called epigenetic theory, put forward by many workers long ago[10,163,164] is about to stage a comeback!

IV. CONCLUSIONS

There is no doubt that many of the chemical carcinogenesis systems that have evolved do pose problems so far as biochemical studies of oncogenesis are concerned. Much of the earlier data should be treated with caution as the extensive cell damage and death induced by some carcinogenesis regimens tends to cloud the true nature of neoplastic change. There are now carcinogenesis systems available, which when used with care, do not involve these disadvantages. Use of these systems should enable us to obtain an accurate picture of the biochemical nature of carcinogenesis.

To summarize, the following factors must be taken into consideration if meaningful experiments are to be carried out:

1.   The dosage of the carcinogen employed should be such that extreme cell damage and death does not occur in the target tissue to be examined. The investigators should conduct their own independent pathological and histological examinations on the carcinogenesis system they propose to use for biochemical studies. In this connection it is as well to ascertain that the dose regimens *do* actually give tumors (as reported by other investigators).
2.   Some estimate of the transformation efficiency in the target organ should be obtained. Obviously if only a few cells (clonal), or a selected cell type, is transformed, then attempts should be made to isolate the affected area prior to biochemical studies. Alternatively, it would be better to start with a system in which the target organ is large enough to permit the isolation of sufficient material for analysis: naturally the more homogeneous the cell population the better.
3.   Planning of the logistics of a long-term carcinogenesis experiment should be done with care at the onset. It is important to use sufficient animals to allow for biological variation and for health problems, deaths, etc.
4.   Attention should be paid to the safety of the operatives involved — can the carcinogen be safely handled with the facilities available? With some of the more potent carcinogens, such as aflatoxins and nitrosamines, special precautions have to be taken.
5.   Sex differences in carcinogenic response should be taken into consideration; for example, females are not as susceptible to urethane carcinogenesis as males.
6.   Comparision of tumor material with the tissue of origin is unlikely to tell us much about the process of oncogenesis.

V. SUGGESTIONS FOR FUTURE WORK

Additionally, the author would strongly advise those working on the effects of chemical carcinogens in the genetic material to consider the following important factors.

1.   Good methods should be available for the preparation of clean nuclei in high yield from target tissue. This is often difficult to attain. For example, we have found that some classes of nuclei become more fragile in carcinogen-treated tissue and are destroyed during the fractionation procedure.[173] Also the presence of detergents is often necessary to achieve good nuclear yields but, they release proteolytic enzymes from the lysozomes and activate the internal nuclear proteases. It is therefore important to include protease inhibitors throughout these preparations.

2. Wherever possible attempts should be made to separate out different types of cell nuclei, for example, as has been done in the case of liver nuclei.[31,149]
3. It would be highly desirable to develop suitable methods to separate "active" from "inactive" chromatin fractions. Methods are available but each has its merits and disadvantages.
4. Some standardized methods of nuclear protein extraction should be used, thus making it easier to compare the results from different laboratories. There are many pitfalls involving nuclear protein extraction (e.g., see the chapter written by MacGillivray[16]). Incomplete extraction methods delude investigators into thinking they have found quantitative and qualitative changes in nuclear protein analytical patterns.
5. Single-dimensional separations of NHP are of limited use. Now that two-dimensional systems with high resolving power are available these should be used to identify the species of NHP present.[150,165-167]
6. The investigator should be able to demonstrate clearly that the results obtained are reproducible. With the large number of manipulations and steps involved in a carcinogenesis experiment, reproducibility of data with each run is often difficult to demonstrate. Material should not be submitted for publication until this is done. The NHP are notoriously unstable moieties and artifactual results are commonplace in the literature. Inexperienced investigators should take advice before entering the field!

Unfortunately, if many of the points mentioned above are adhered to, other problems can arise. For example, the more efficient a carcinogen, the greater its toxicity; also the lower dose regimens require long-term administration of carcinogens to be made.

In our experience very few of the "happy intermediate" systems are available to the carcinogenesis researcher. We have tried "one-dose" carcinogenesis systems but often find tissue damage in the target organs (by histological examination) to be unacceptably high. We have also tried low dose systems only to find that the incidence of transformed cells can be too low. How then should one approach future in vivo carcinogenesis systems? The answer almost certainly lies in the newer type of carcinogenesis systems being developed, in which phased steps in the oncogenic process are used. These can be compared to older studies such as those classically done on skin (e.g., with croton oil and polycyclic aromatic hydrocarbons) where initiation and promotion stages have been described over a prolonged time course. Unfortunately, the skin system does not lend itself to biochemical studies on nuclear proteins. However, liver carcinogenesis can now be examined, using a program of similar progressive steps in which the sequential events involved can be studied. These regimens involve combinations of carcinogens, noncarcinogen "promoters", growth stimulators, or, stimulation such as partial hepatectomy. Tumors are not normally produced by treatment with the individual agents alone but only by combinations of them.

These regimens usually involve the formation of islands of cancer-"initiated" cells in the livers of the treated animals. These "initiated" cells are not malignant and cannot be detected by morphological means, but can histochemically, as enzyme altered foci (EAF). These initiated cells can then be induced to progress slowly ("promoted") to tumor cell phenotypes (malignant hepatocellular carcinoma). Thus, a sequential analysis of events can be performed allowing the investigator to study early putative preneoplastic stages.

For studies on the biochemistry of hepatocarcinogenesis the following four systems can be recommended:

System 1. First suggested by Peraino and associates[168]. Young animals are given a relatively subcarcinogenic dose (or doses) of AAF. Liver tumors are then induced by the administration of phenobarbital (the latter is not capable of inducing tumors in a significant fashion when used alone).

Regimen: 1. AAF for 3 weeks
   2. Basal diet for 1 week
   3. 0.05% phenobarbital over 5 months

This treatment gives 100% incidence of hepatomas.

System 2. Solt and Farber[169] have developed a new approach to the sequential analysis of carcinogenesis that they claim delineates the first steps in the process and can be used for a quantitative assay for the initiation of cancer.

Regimen: 1. DEN (other carcinogens are also effective) daily for 1 week, dose is maximized for foci production (50 foci per cm²)
   2. Basal diet-recovery — 2 weeks
   3. 0.02% AAF in diet — 1 week
   4. Partial hepatectomy (67%)

System 3. As described by Pitot and co-workers.[170,171]

Regimen: 1. Initiation of foci by a single dose of DEN following partial hepatectomy (24 hr)
   2. Normal basic diet for 8 weeks
   3. "Promotion" by continuous feeding of 0.05% phenobarbital in the diet for 24 weeks

After the first traumatic treatments the liver regenerates and settles down; EAF are formed.

System 4. As recently described by Cayama et al.[172]

Regimen: 1. N-methyl-N-nitrosourea (NMU) given IP (not a liver carcinogen)
   2. 3 hr later, when all NMU used up — partial
   hepatectomy
   3. Recovery period on basic diet, 2 weeks
   4. 0.02% AAF administered in diet for 1 week
   5. α-hexachlorocyclohexane (a liver mitogen given to stimulate cell proliferation (and therefore appearance of tumors)

If all the factors mentioned here are taken into account and controlled experiments are carried out on the type of chemical carcinogenesis systems described above, it should not be long before a basic understanding of the process of oncogenesis is within our grasp.

ACKNOWLEDGMENTS

I wish to thank Fraser Lewis and Tony Thackrah for all their invaluable help during my efforts to tackle the nonhistone problem in relation to cancer. Also, to Mrs. Margaret Harris for typing and Mr. Brian Pentecost for help in the preparation of this manuscript.

REFERENCES

1. Clayson, D. B., *Chemical Carcinogenesis,* J & A Churchill, London, 1962.
2. Goldberg, L., Ed., *Carcinogenesis Testing of Chemicals,* CRC Press, Boca Raton, Fla., 1974.
3. Searle, C. E., Ed., *Chemical Carcinogenesis,* Monograph No. 173, American Chemical Society, Washington, D.C., 1976.
4. Becker, F. E., Sequential phenotypic and biochemical alterations during chemical hepatocarcinogenesis, *Cancer Res.,* 36, 2563, 1976.
5. Miller, E. C., The presence and significance of bound aminoazo dyes in the liver of rats fed p-DAB, *Cancer Res.,* 7, 468, 1947.
6. Sorof, S., Young, E. M., and Otto, M. G., Soluble liver h proteins during hepatocarcinogenesis by aminoazo dyes and 2-acetylaminofluorene in the rat, *Cancer Res.,* 18, 32, 1958.
7. Abell, C. W. and Heidelberger, C., Interactions of carcinogenic hydrocarbons with tissue VIII. Binding of tritium labelled hydrocarbons to the soluble proteins of mouse skin, *Cancer Res.,* 22, 931, 1962.
8. Miller, J. A. and Miller, E. C., The metabolic activation of carcinogenic aromatic amines and amides, in *Progress in Experimental Tumour Research,* Vol. 11, S. Karger, Basel, 1969. 273.
9. Miller, E. C. and Miller, J. A., *Molecular Biology of Cancer,* Busch, H., Ed., Academic Press, New York, 1974, 377.
10. Heidelberger, C., Chemical carcinogenesis, *Annu. Rev. Biochem.,* 44, 79, 1975.
11. Ketterer, B., Tipping, E., Beale, D., Menwissen, J., and Kay, C. M., Proteins which specifically bind carcinogens, in *Biochem. 11th Interim Cancer Congr.,* Vol. 2, Chemical and Viral Carcinogenesis, Bucalossi, P., Veronesi, U., and Cascinelli, N., Eds., Excerpta Medica, Amsterdam, 1975, 25.
12. Irving, C. C., Interaction of chemical carcinogenesis with DNA, in *Methods in Cancer Research,* Vol. 7, Busch, H., Ed., Academic Press, New York, 1973, 189.
13. Ames, B. N. and Hooper, K., Does carcinogenic potency correlate with mutagenic potency in the Ames assay?, *Nature (London),* 274, 19, 1978.
14. Hart, R. W., Setlow, R. B., and Woodhead, A. D., Evidence that pyrimidine dimers in DNA can give rise to tumors, *Proc. Natl. Acad. Sci. U.S.A.,* 74, 5574, 1977.
15. Gurdon, J. B. and Uehlinger, V., "Fertile" intestine nuclei, *Nature (London),* 210, 1240, 1966.
16. Birnie, G. D., Ed., *Subnuclear Components,* Butterworths, London, 1976.
17. Pegg, A. E., Alkylation of rat liver DNA by DMS, effect of dosage on O^6-methylguanine levels, *J. Natl. Cancer Inst.,* 54, 681, 1977.
18. Margison, C. P., Margison, J. M., and Montesano, R., Accumulation of O^6-methylguanine in non-target-tissue DNA during chronic administration of DMN, *Biochem. J.,* 165, 463, 1977.
19. Scherer, E., Steward, A. P., and Emmelot, P., Kinetics of formation of O^6-methylguanine in, and its removal from, liver DNA of rats receiving diethylnitrosamine, *Chem. Biol. Interact.,* 19, 1, 1977.
20. Huberman, E., Mager, R., and Sachs, L., Mutagenesis and transformation of normal cells by chemical carcinogens, *Nature (London),* 264, 360, 1976.
21. Ts'o, P. O. P., Ed., *The Molecular Biology of the Mammalian Genetic Apparatus,* Vols. 1 and 2, North-Holland, Amsterdam, 1977, 241.
22. Allfrey, V. G. and Boffa, L. C., Modifications of nuclear protein structure and function during carcinogenesis, in *The Cell Nucleus,* Vol. 7, Busch, H., Ed., Academic Press, New York, 1979, 521.
23. Gronow, M., Nuclear proteins and chemical carcinogenesis, *Chem. Biol. Interact.,* 29, 1, 1980.
24. Kornberg, R., The location of nucleosomes in chromatin: specific or statistical?, *Nature (London),* 292, 579, 1981.
25. Miller, J. A., Carcinogenesis by chemicals: an overview. G. H. A. Clowes Memorial Lecture, *Cancer Res.,* 30, 559, 1970.
26. Kriek, E., Carcinogenesis by aromatic amines, *Biochim. Biophys. Acta,* 355, 177, 1974.
27. Barry, E. J., Ovechka, C. A., and Gutmann, H. T., Interaction of aromatic mines with rat liver protein in vivo. II. Binding of N-2-fluoreneylacetamide 9-^{14}C to nuclear proteins, *J. Biol. Chem.,* 243, 51, 1968.
28. Lotlikar, P. D. and Paik, W. M., Binding of carcinogenic aromatic amine to rat liver, nuclear acidic proteins in vivo, *Biochem. J.,* 124, 443, 1971.
29. Stein, G. S., Stein, J. A., and Thomson, J. A., Chromosomal proteins in transformed and neoplastic cells: a review, *Cancer Res.,* 38, 1181, 1978.
30. Gronow, M. and Lincoln, J. C., The binding of ^{14}C-labelled acetylaminofluorene (AAF) to nuclear proteins and DNA, *Cancer Lett.,* 5, 269, 1978.
31. Tulp, A., Welagen, J. J. M. N., and Westra, J. G., Binding of the chemical carcinogen N-Hydroxy-acetyl-aminofluorene to ploidy classes of rat liver nuclei as separated by velocity sedimentation at unit gravity, *Chem. Biol. Interact.,* 23, 293, 1978.
32. Schwartz, E. L. and Goodman, J. I., Quantitative and qualitative aspects of the binding of N-hydroxy-2-acetylaminofluorene to hepatic chromatin fractions, *Chem. Biol. Interact.,* 26, 287, 1979.

33. Metzger, G. and Werbin, H., Evidence for N-acetoxy-N-2-acetylaminofluorene induced covalent-like binding of some non-histone proteins to DNA in chromatin, *Biochemistry*, 18, 655, 1979.

34. Stout, D. L., Hemminki, K., Becker, F. F., Covalent binding of 2-acetyl-aminofluorene, 2-aminofluorene, and N-hydroxy-2-acetylaminofluorene to rat liver nuclear DNA, *Cancer Res.*, 40, 3579, 1980.

35. Zieve, F. J., Inhibition of rat liver RNA polymerase by the carcinogen N-hydroxy-2-fluoreneacetamide, *Biol. Chem.*, 247, 5987, 1972.

36. Glazer, R. I., Nutter, R. C., Glass, L. E., and Menger, F. M., 2AAF and N-hydroxy 2AAF inhibition of incorporation of orotic acid-5-³H into nuclear ribosomal and heterogeneous RNA in normal and regenerating liver, *Cancer Res.*, 34, 2451, 1974.

37. Glazer, R. I., Glass, L. E., and Menger, F. M., Modification of hepatic RNA polymerase activities by N-hydroxy 2-AAF and N-acetoxy-2 AAF, *Mol. Pharmacol.*, 11, 36, 1975.

38. Herzog, J., Serroni, A., Briesmaster, B. A., and Farber, J. L., N-Hydroxy-2-acetylaminofluorene inhibition of rat liver RNA polymerases, *Cancer Res.*, 35, 2138, 1975.

39. Glazer, R. I., Alterations in the transcriptional capacity of hepatic DNA-dependent RNA polymerase I and II by N-hydroxy-2-acetylaminofluorene, *Cancer Res.*, 36, 2282, 1976.

40. Craddock, V. M., DNA polymerases in replication and repair of DNA during carcinogenesis induced by feeding N-acetylaminofluorenes, *Carcinogenesis*, 2, 61, 1981.

41. Wogan, G. N., Aflatoxin carcinogenesis, in *Methods in Cancer Research*, Vol. 7, Busch, H., Ed., Academic Press, New York, 1973, 309.

42. Vaught, J. B., Klohs, W., and Gurtoo, H. L., In vitro metabolism of Aflatoxin B1 by rat liver nuclei, *Life Sci.*, 21, 1497, 1977.

43. Lafarge, C. and Frayssinet, C., The reversibility of inhibition of RNA and DNA synthesis induced by aflatoxin in rat liver. A tentative explanation for carcinogenic mechanism, *Int. J. Cancer*, 6, 74, 1970.

44. Pong, R. S. and Wogan, G. N., Time course and dose-response characteristics of aflatoxin; effects on rat liver RNA polymerase and ultrastructure, *Cancer Res.*, 30, 294, 1970.

45. Groopman, J. D., Bushy, W. F., Jr., and Wogan, G. N., Nuclear distribution of aflatoxin B1 and its interaction with histones in rat liver *in vivo*, *Cancer Res.*, 40, 4343, 1980.

46. Morris, H. P., Hepatocarcinogenesis by 2-acetylaminofluorene and related compounds including comments on dietary and other influences, *Natl. Cancer Inst.*, 15, 1535, 1955.

47. Miller, E. C. and Miller, J. A., The carcinogenic azo-dyes, *Adv. Cancer Res.*, 1, 339, 1953.

48. Kadlibar, F. F., Miller, J. A., and Miller, E. C., Microsomal N-oxidation of the hepatocarcinogen N-methyl-4-aminoazobenzene and the reactivity of N-hydroxy-N-methyl-4aminoazobenzene, *Cancer Res.*, 36, 1196, 1976.

49. Terayama, H., Aminoazo carcinogenesis-methods and biochemical problems, in *Methods in Cancer Research*, Busch, H., Ed., Academic Press, New York, 1967, 399.

50. Miller, E. C. and Miller, J. A., The presence and significance of bound aminoazo dyes in the liver of rats fed p-dimethylaminoazobenzene, *Cancer Res.*, 7, 468, 1947.

51. Gelboin, H. V., Miller, J. A., and Miller, E. C., Studies on hepatic protein-bound dye formation in rats given single large doses of 3'MeDAB, *Cancer Res.*, 18, 608, 1958.

52. Bakay, B. and Sorof, S., Soluble nuclear proteins of liver and tumour in azo dye carcinogenesis, *Cancer Res.*, 24, 1814, 1964.

53. Scribine, J. D., Miller, J. A., and Miller, E. C., 3-methyl-mercapto-N-methyl-4-aminoazobenzene. An alkaline degradation product of a labile protein bound dye in the livers of rats fed N,N dimethyl-4-aminoazobenzene, *Biochem. Biophys. Res. Commun.*, 20, 560, 1965.

54. Rees, K. R., Rowland, G. F., and Varcoe, J. S., Studies on the binding of tritiated p-dimethyl-aminoazobenzene in rat liver and the role of intranuclear binding sites in the early stages of carcinogenesis, *Br. J. Cancer*, 19, 903, 1965.

55. Rees, K. R. and Varcoe, J. S., The interaction of tritiated-p-dimethyl-aminoazobenzene with rat liver nuclear proteins, *Br. J. Cancer*, 21, 174, 1967.

56. Roberts, J. J. and Warwick, G. R., The covalent binding of metabolites of dimethylaminoazobenzene, β-naphthylamine and aniline to nucleic acids in vivo, *Int. Cancer*, 1, 179, 1966.

57. Dijkstra, J. and Griggs, H. M., Binding of aminoazo dyes to serum albumin and to nuclear proteins of rat liver, *Br. J. Cancer*, 21, 205, 1967.

58. Bakay, B. and Sorof, S., Zonal electrophoresis of the soluble nuclear proteins of normal and preneoplastic livers, *Cancer Res.*, 29, 22, 1969.

59. Albert, A. E., Studies of the interactions of aromatic amino azo compounds with macromolecules, Ph.D. dissertation, University of London, 1970.

60. Albert, A. E. and Warwick, G. P., The intrachromosomal distribution of ³H-dimethylaminoazobenzene in rat nuclei in vivo, *Chem. Biol. Interact.*, 5, 61, 1972.

61. Albert, A. E. and Warwick, G. P., The subcellular distribution of tritiated 4-dimethylaminoazobenzene and 2-methyl-4-dimethylaminoazobenzene in rat liver and spleen following a single oral administration, *Chem. Biol. Interact.*, 5, 65, 1972.

62. Jungmann, R. A. and Schweppe, J. S., Binding of chemical carcinogens to nuclear proteins of rat liver, *Cancer Res.*, 32, 952, 1972.
63. Chauveau, J., Meunier, N., and Benoit, A., Binding of metabolites of dietary 4-dimethylaminoazobenzene and 2-methyl-4-dimethylaminoazobenzene to rat liver DNA and protein of subcellular fractions, *Int. J. Cancer*, 13, 1, 1974.
64. Yoshida, M. and Holoubek, V., Binding of 3-methyl-4-dimethylaminoazobenzene to nuclear ribonucleoprotein particles, *Life Sci.*, 18, 49, 1976.
65. Sonnenbichler, J. and Reichart, F., Wechselwirkung von p-domethylaminoazobenzol mit nichthistonprotein aus rattenleber-chromatin, *Z. Krebsforsch.*, 91, 55, 1978.
66. Akao, M., Kuroda, K., and Miyaki, K., Selective inhibition of RNA polymerase activity in rat liver nuclei by 4-(dimethylamino) azobenzene and the effect of nitrofurans on rat liver RNA metabolism associated with prevention of carcinogenesis, *Gann*, 63, 1, 1972.
67. Akao, M., Kuroda, K., Tsutsui, Y., Kanisawa, M., and Miyoki, K., Effect of nitrofurans antagonistic to 3'-methyl-4-dimethylaminoazobenzene in hepatocarcinogenesis and RNA polymerase activity, *Cancer Res.*, 34, 1843, 1974.
68. Chiu, J. F., Craddock, C., and Hnilica, L. S., Bound forms of nuclear DNA polymerase in regenerating rat livers, *FEBS Lett.*, 36, 235, 1973.
69. Chiu, J. F., Craddock, C., Morris, H. P., and Hnilica, L. S., Nuclear DNA polymerases during experimental neoplasia of rat liver, *Cancer Biochem. Biophys.*, 1, 13, 1974.
70. Phillips, W. A. and Blunck, J. M., Inhibition of chemical carcinogenesis, increased activity of soluble RNA polymerase EC-2.7.7.6 in the liver of rats protected against 3'-methyl-4-dimethylamino azobenzene hepatocarcinogenesis by dietary chloramphenicol, *Eur. J. Cancer*, 13, 729, 1977.
71. Rees, K. R. and Rowland, G. F., The metabolism of isolated rat liver nuclei during chemical carcinogenesis, *Biochem. J.*, 80, 428, 1960.
72. Saharabudhe, M. B., Apte, B. K., Aboobaker, V. S., and Jayaraman, R., Partial deletion of aspartic acid from DNA-proteins during butter yellow carcinogenesis, *Biochem. Biophys. Res. Commun.*, 7, 173, 1962.
73. Hawtrey, A. O. and Nourse, L. D., The effect of 3'-MeDAB on protein synthesis and DNA dependent RNA polymerase activity in rat liver nuclei, *Biochim. Biophys. Acta*, 80, 530, 1964.
74. Sporn, M. B. and Dingman, C. W., Effects of carcinogens and hormones on rat liver chromatin, *Cancer Res.*, 26, 2488, 1966.
75. Dijkstra, J. and Weide, S. S., Changes in chromatin caused by aminoazo compounds, *Exp. Cell Res.*, 72, 1972, 345, 1972.
76. Albert, A. E., Alterations in turnover of labelled precursors incorporated into rat liver nuclear macromolecules during azo dye feeding, *Chem. Biol. Interact.*, 4, 287, 1972.
77. Yoshida, M. and Holoubek, V., Early effects of carcinogenic aminoazo dyes on the protein patterns and metabolism in rat liver, *Int. J. Biochem.*, 7, 259, 1976.
78. Patel, N. T. and Holoubek, V., Protein composition of liver nuclear ribonucleoprotein particles of rats fed carcinogenic aminoazo dyes, *Biochem. Biophys. Res. Commun.*, 73, 112, 1976.
79. Patel, N. T. and Holoubek, V., Protein composition of nuclear ribonucleoprotein particles isolated from liver of rats in the early stages of feeding of 3-methyl-4-dimethylaminoazobenzene and from hepatoma induced by the same carcinogen, *Chem. Biol. Interact.*, 19, 303, 1977.
80. Chiu, J. F., Craddock, C., Getz, S., and Hnilica, L. S., Non-histone chromatin protein phosphorylation during azo-dye carcinogenesis, *FEBS Lett.*, 33, 247, 1973.
81. Ahmed, K., Increased phosphorylation of nuclear phosphoproteins in precancerous liver, *Res. Commun. Chem. Pathol. Pharmacol.*, 9, 771, 1974.
82. Chiu, J. F., Brade, W. P., Thompson, J., Tsai, Y. H., and Hnilica, L. S., Non-histone protein phosphorylation in normal and neoplastic rat liver chromatin, *Exp. Cell. Res.*, 91, 200, 1975.
83. Chiu, J. F., Hunt, M., and Hnilica, L. S., Tissue-specific DNA-protein complex during azo dye hepatocarcinogenesis, *Cancer Res.*, 35, 913, 1975.
84. Allfrey, V. G., Boffa, L. C., and Vidali, G., Changes in composition and metabolism of nuclear non-histone proteins during chemical carcinogenesis. Association of tumour-specific DNA-binding proteins with DNAase I-sensitive regions of chromatin and observations on the effects of histone-acetylation on chromatin structure, *Miami Winter Symposia*, Vol. 15, Differentiation and Development, Ahmad, F., Russell, T. R., Schultz, J., and Werner, R., Eds., Academic Press, New York, 1978, 261.
85. Allfrey, V. G. and Boffa, L. C., Changes in chromosomal protein in colon cancer, *Cancer (Philadelphia)*, 40, 2576 and 2584, 1977.
86. Boffa, L. C., Vidali, G., and Allfrey, V. G., Changes in nuclear NHP composition during normal differentiation and carcinogenesis of intestinal epithelial cells, *Exp. Cell Res.*, 98, 396, 1976.
87. Boffa, L. C., Vidali, G., and Allfrey, V. G., Selective synthesis and accumulation of nuclear non-histone proteins during carcinogenesis of the colon induced by 1, 2-dimethylhydrazine, *Cancer*, 36, 2356, 1975.

88. Boffa, L. C. and Allfrey, V. G., Characteristic complements of nuclear non-histone proteins in colonic epithelial tumours, *Cancer Res.*, 36, 2678, 1976.
89. Boffa, L. C., Karn, J., Vidali, G., and Allfrey, V. G., Distribution of N^G, N^G, dimethylarginine in nuclear protein factions, *Biochem. Biophys. Res. Commun.*, 74, 969, 1977.
90. Boffa, L. C. Diwan, B. A., Gruss, R., and Allfrey, V. G., Differences in colonic nuclear proteins of two mouse strains with different susceptibilities to 1,2-dimethylhydrazine-induced carcinogenesis, *Cancer Res.*, 40, 1774, 1980.
91. Chiu, J. F., Pechaumphai, W. Markert, C., and Little, D. W., Immunospecificity of nuclear non-histone protein-DNA complexes in colon adenocarcinoma, *J. Natl. Cancer Inst.*, 63, 313, 1979.
92. Chiu, J. F., Pumo, D., and Gootnick, D., Antigenic changes in nuclear chromatin during 1,2-dimethylhydrazine-induced colon carcinogenesis, *Cancer Res,*. 45(Suppl. 5), 1193, 1980.
93. Farber, E., Ethionine carcinogenesis, *Adv. Cancer Res.*, 7, 383, 1963.
94. Stekol, J. A., Biochemical basis for ethionine effects on tissues, *Adv. Enzymol.*, 25, 369, 1963.
95. Smuckler, E. A. and Koplitz, M., The effects of carbon tretrachloride and ethionine on RNA synthesis in vivo and isolated rat liver nuclei, *Arch. Biochem. Biophys.*, 132, 62, 1969.
96. Farber, J. L., Shinozuka, H., Serroni, A., and Farmar, R., Reversal of the ethionine-induced inhibition of rat liver ribonucleic acid polymerases in vivo by adenine, *Lab. Invest.*, 31, 465, 1974.
97. Farber, J. L., McConomy, J., Franzen, B., Marroquin, F., Stewart, G. A., and Magee, P. N., Interaction between ethionine and rat liver ribonucleic acid and protein *in vivo*, *Cancer Res.*, 27, 1761, 1967.
98. Friedman, M., Shull, K. H., and Farber, E., Highly selective in vivo ethylation of rat liver nuclear protein by ethionine, *Biochem. Biophys. Res. Commun.*, 34, 857, 1969.
99. Orenstein, J. M. and Marsh, W. H., Incorporation in vivo of methione and ethionine into, and the methylation and ethylation of, rat liver nuclear proteins, *Biochem. J.*, 109, 697, 1968.
100. Wogan, G. N. and Tannenbaum, S. R., Environmental N-nitroso compounds: implications for public health, *Toxicol. Appl. Pharmacol.*, 31, 375, 1975.
101. Druckrey, H., Quantitative aspects in chemical carcinogenesis, in *Potential Hazards from Drugs*, Truhaut, R., Ed., UICC Monograph Vol. 7, Springer-Verlag, Berlin, 1965, 60.
102. Christie, C. S. and LePage, R. N., Enlargement of liver cell nuclei: effect of dimethylnitrosamine on size and desoxyribosenucleic acid content, *Lab. Invest.*, 10, 729, 1961.
103. Hitachi, M., Yamada, K., and Yakayama, S., Diethylnitrosamine-induced chromosome changes in rat liver cells, *J. Natl. Cancer Inst.*, 53, 507, 1974.
104. Stewart, B. W. and Farber, E., Apparent changes in structure of rat liver chromatin following administration of alkylating carcinogens, *Proc. Aust. Biochem. Soc.*, 7, 90, 1974.
105. Nicolini, C., Ramanathan, R., Kendall, F., Murphy, J., Parodi, S., and Sarma, D. S. R., Physico-chemical alterations in the conformaton of rat liver chromatin induced by carcinogens in vivo, *Cancer Res.*, 36, 1725, 1976.
106. Magee, P. N., Nitrosamine carcinogenesis, in *Chemical Carcinogenesis*, Searle, C. E., Ed., Monograph 173, American Chemical Society, Washington, D.C., 1976.
107. Montesano, R. and Bartsch, H., Mutagenic and carcinogenic N-nitroso compounds: possible environmental hazards, *Mutation Res.*, 32, 179, 1976.
108. Swann, P. F. and McLean, A. E. M., Cellular injury and carcinogenesis. The effect of a protein-free higher carbohydrate diet on the metabolism of DNA in the rat, *Biochem. J.*, 124, 283, 1971.
109. Turberville, C. and Craddock, V. M., Methylation of nuclear proteins by dimethylnitrosamine and methionine in the rat in vivo, *Biochem. J.*, 124, 725, 1971.
110. Galbraith, A. and Itzhaki, R., Studies on histones and non-histone proteins from rats treated with dimethylnitrosamine, *Chem. Biol. Interact.*, 28, 309, 1979.
111. Pinsky, S., Tew, K. D., Smulson, M. E., and Woolley, P. V., III, Modification of L 1210 cell nuclear proteins by 1-methyl-1-nitrosourea and 1-methyl-3-nitro-1-nitrosoguanidine, *Cancer Res.*, 39, 923, 1979.
112. Zytkovicz, T. H., Moses, H. L., and Spelsberg, T. C., Nuclear interactions of polycyclic aromatic hydrocarbons, in *The Cell Nucleus*, Busch, H., Ed., Vol. 7, Academic Press, New York, 1979, 479.
113. Herzog, J. and Farber, J. L., Inhibition of rat liver RNA polymerases by action of methylating agents dimethylnitrosamine in (vivo) and methyl methamesulphonate in vitro, *Cancer Res.*, 36, 1761, 1976.
114. Craddock, V. M., Cell proliferation and experimental liver cancer, in *Liver Cell Cancer*, Linsell, C. A. and Warwick, G. P., Eds., Elsevier, Amsterdam, 1976, 153.
115. Stewart, B. W. and Farber, E., Alterations in thermal stability of rat liver chromatin and DNA induced in vivo by dimethylnitrosamine and diethylnitrosamine, *Cancer Res.*, 38, 510, 1978.
116. Craddock, V. M. and Ansley, C. M., Sequential changes in DNA polymerases α and β during diethylnitrosamine-induced carcinogenesis, *Biochim. Biophys. Acta*, 564, 15, 1979.
117. Berthold, V., Thielmann, H. W., and Geider, K., Carcinogens inhibit DNA synthesis with isolated DNA polymerases from *Escherichia coli*, *FEBS Lett.*, 86, 81, 1978.

118. Gronow, M. and Griffiths, G., Changes in rat liver nuclear protein metabolism following a single carcinogenic dose of diethylnitrosamine, *Exp. Pathol.*, 9, 73, 1974.

119. Gronow, M. and Thackrah, T. M., Nuclear protein changes during the nitrosamine-induced carcinogenesis of rat liver, *Chem. Biol. Interact.*, 9, 225, 1974.

120. Cronow, M. and Thackrah, T. M., Changes in the composition of rat liver chromatin fractions during nitrosamine carcinogenesis, *Eur. J. Cancer*, 10, 21, 1974.

121. Alonso, A. and Arnold, H. P., Stimulation of amino acids incorporation into rat liver nonhistone chromatin proteins after treatment with diethyl nitrosamine, *FEBS Lett.*, 41, 8, 1974.

122. Augenlicht, L. H., Biessmann, H., and Rajewsky, M. F., Chromosomal proteins of rat brain: increased synthesis and affinity for DNA following a pulse of carcinogen ethylnitrosourea in vivo, *J. Cell Physiol.*, 86, 431, 1975.

123. Tsanev, R. and Hadjiolov, D., Chromosomal proteins in hepatocarcinogenesis, *Z. Krebsforsch.*, 91, 237, 1978.

124. Craddock, V. M. and Henderson, A. R., The effect of dimethylnitrosamine on the synthesis of high-mobility group non-histone proteins in regenerating rat liver, *Carcinogenesis*, 1, 445, 1980.

125. Martinez-Sales, V., Gabaldon, M., and Baguena, J., Nonhistone protein changes during the diethylnitrosamine induced carcinogenesis of rat liver, *Cancer Res.*, 41, 1187, 1981.

126. Letnansky, K., The phosphorylation of nuclear proteins in the regenerating and premalignant rat liver and its significance for cell proliferation, *Cell Tissue Kinet.*, 8, 423, 1975.

127. Brookes, P., Covalent interactions of carcinogens with DNA, *Life Sci.*, 16, 331, 1975.

128. Brookes, P. and Lawley, P. D., Evidence for the binding of polycyclic aromatic hydrocarbons to the nucleic acids of mouse skin, *Nature (London)*, 202, 781, 1964.

129. Sims, P. and Grover, P. L., Epoxides in polycyclic aromatic hydrocarbon metabolism and carcinogenesis, *Adv. Cancer Res.*, 20, 165, 1974.

130. Neidle, S., Carcinogen binding to DNA, *Nature (London)*, 276, 444, 1978.

131. O'Brien, R. L., Stauton, R., and Craig, R. L., Chromatin binding of benzo (a) pyrene and 20 methylcholanthrene, *Biochim. Biophys. Acta*, 186, 414, 1969.

132. Vaught, J. and Bresnick, E., Binding of polycyclic hydrocarbons to nuclear components in vitro, *Biochem. Biophys. Res. Commun.*, 69, 587, 1976.

133. Bresnick, E., Vaught, J. B., Chuang, A. H. C., Stroming, T. A., Bockman, D., and Mukhtar, H., Nuclear aryl hydrocarbon hydroxylase and in interaction of PAH with nuclear components, *Arch. Biochem. Biophys.*, 181, 257, 1977.

134. Pezzuto, J. M., Lea, M. A., and Yang, C. S., The role of microsomes and nuclear envelope in the metabolic activation of benzo (a) pyrene leading to binding with nuclear macromolecules, *Cancer Res.*, 37, 3427, 1977.

135. Zytkovicz, T. H., Moses, H. L., and Spelsberg, T. C., Nuclear interaction of polycyclic aromatic hydrocarbons, in *The Cell Nucleus*, Busch, H., Ed., Vol. 7, Academic Press, New York, 1979, 479.

136. Jahn, C. L. and Litman, G. W., Distribution of covalently bound benzo (a) pyrene in chromatin, *Biochem. Biophys. Res. Commun.*, 76, 534, 1977.

137. Jahn, C. L. and Litman, G. W., Accessibility of DNA in chromatin to the covalent binding of the chemical carcinogen. Benzo[a] pyrene (BP), *Biochemistry*, 18, 1442, 1979.

138. Kootstra, A., Shah, Y. B., and Slaga, T. J., Binding of Benzo [a] pyrene diol-epoxide (anti) to nucleosomes containing high mobility group proteins, *FEBS Lett.*, 116, 62, 1980.

139. Kootstra, A., Haas, B. L., and Slaga, T. J., Reactions of benzo [a] pyrene diol-epoxides with DNA and nucleosomes in aqueous solutions, *Biochem. Biophys. Res. Commun.*, 94, 1432, 1980.

140. MacLeod, M. C., Mansfield, B. K., Huff, A., and Selkirk, J. K., Simultaneous preparation of nuclear DNA, RNA and protein from carcinogen treated hamster embryo fibroblasts, *Anal.* Biochem., 97, 410, 1979.

141. MacLeod, M. C., Kootstra, A., Mansfield, B. K., Slaga, T. J., and Selkirk, J. K., Specificity in interaction of benzo [a] pyrene with nuclear macromolecules: implication of derivatives of two dihydrodiols in protein binding, *Proc. Natl. Acad. Sci. U.S.A.*, 77, 6396, 1980.

142. Jerina, D. M. et al., *Origins of Human Cancer*, Hiatt, H. H., Watson, J. D., and Weinstein, J. A., Eds., Vol. 4, Cold Spring Harbor Laboratory, Cold Spring Harbor, New York, 1977, 639.

143. Rogan, E. G., Mailander, P., and Cavalieri, E., Metabolic activation of aromatic hydrocarbons in purified rat liver nuclei: induction of enzyme activities and binding to DNA with and without monooxygenase-catalyzed formation of active oxygen, *Proc. Nat. Acad. Sci. U.S.A.*, 73, 457, 1976.

144. Adams, H. R. and Busch, H., Labelling of nuclear RNA with orotic acid-2-^{14}C in livers of normal and thioacetamide treated rats, *Biochem. Biophys. Res. Commun.*, 9, 578, 1962.

145. Villalobos, J. G., Steele, W. J., and Busch, H., Ribonuclease activity of isolated nucleoli of livers of thioacetamide-treated rats, *Biochim. Biophys. Acta*, 103, 195, 1965.

146. Barton, A. D., and Anderson, E. C., Effects of administration of thioacetamide on chromosome associated proteins of rat liver, *Arch. Biochem. Biophys.*, 111, 206, 1965.

147. Barton, A. D. and Anderson, E. C., Observations on the histone fraction from rat liver and its alteration in animals treated with thioacetamide, *Arch. Biochem. Biophys.*, 117, 471, 1966.

148. **Muramatsu, M. and Busch, H.,** Effects of thioacetamide on metabolism of proteins of normal and regenerating liver, *Cancer Res.,* 22, 1100, 1962.
149. **Gonzales-Mujica, F. and Mathias, A. P.,** Proteins from different classes of liver nuclei in normal and thioacetamide-treated rats, *Biochem. J.,* 133, 441, 1973.
150. **Yeoman, L. C., Taylor, C. W., and Busch, H.,** 2-D gel electrophoresis of acid extractable nuclear proteins of regenerating and thioacetamide treated rat liver, Morris 9618 A hepatoma and Walker 256 carcinoma, *Cancer Res.,* 34, 424, 1974.
151. **Ballal, N. R., Goldknopf, I. L., Goldberg, D. A., and Busch, H.,** The dynamic state of liver nucleolar proteins as reflected by their changes during administration of thioacetamide, *Life Sci.,* 14, 1835, 1974.
152. **Raw, I. and Rockwell, P.,** A protein bound from thioacetamide in liver nucleoli, *Biochem. Biophys. Res. Commun.,* 90, 721, 1979.
153. **Leonard, T. B. and Jacob, S. T.,** Alterations in DNA-dependent RNA polymerase I and II from rat liver by thioacetamide: preferential increase in the level of chromatin-associated nucleolar RNA polymerase IB, *Biochemistry,* 16, 4538, 1977.
154. **Clawson, G. A., Woo, C. H., and Smuckler, E. A.,** Independent response of nucleoside triphosphatase and protein kinase activities in the nuclear envelope following thioacetamide treatment, *Biochem. Biophys. Res. Commun.,* 95, 1200, 1980.
155. **Clawson, G. A., Woo, C. H., and Smuckler, E. A.,** Polypeptide composition of nuclear envelope following thioacetamide-induced nuclear swelling, *Biochem. Biophys. Res. Commun.,* 96, 370, 1980.
156. **Mirvish, S. S.,** The carcinogenic action and metabolism of urethane and N-hydroxyurethane, *Adv. Cancer Res.,* 11, 1, 1968.
157. **Gronow, M. and Lewis, F. A.,** Changes in liver nuclear protein metabolism after a single dose of urethane to suckling mice, *Chem. Biol. Interact.,* 19, 327, 1977.
158. **Prodi, G., Rocchi, P., and Grilli, S.,** In vivo interaction of urethane with nucleic acids and proteins, *Cancer Res.,* 30, 2887, 1970.
159. **Srinivasan, P. R. and Borek, E.,** Enzymatic alteration of nucleic acid structure, *Science,* 145, 548, 1964.
160. **Gantt, R. R.,** Is aberrant nucleic acid methylation related to malignant transformation?, *J. Natl. Cancer Inst.,* 53, 1505, 1974.
161. **Gronow, M. and Thackrah, T. M.,** Alkaline hydrolysis of nuclear proteins, *Biochem. Soc. Trans.,* 7, 196, 1979.
162. **Gronow, M., Lewis, F. A., and Thackrah, T. M.,** Studies on the degradation of HeLa non-histone proteins, *Biochim. Biophys. Acta,* 606, 157, 1980.
163. **Pitot, H. and Heidelberger, C.,** Metabolic regulatory circuits and carcinogenesis, *Cancer Res.,* 23, 1694, 1963.
164. **Farber, E.,** Carcinogenesis — Cellular evolution as a unifying thread, *Cancer Res.,* 33, 2537, 1973.
165. **Barrett, T. and Gould, H. J.,** Tissue and species specificity of nonhistone chromatin proteins, *Biochim. Biophys. Acta,* 294, 165, 1973.
166. **O'Farrell, P. H.,** High resolution two-dimensional electrophoresis of proteins, *J. Biol. Chem.,* 250, 4007, 1975.
167. **Liew, C. C. and Chan, P. K.,** Identification of nonhistone chromatin proteins in chromatin subunits, *Proc. Natl. Acad. Sci. U.S.A.,* 73, 3458, 1976.
168. **Peraino, C., Fry, R. J. M., Staffeldt, E., and Kisieleski, W. E.,** Effects of varying the exposure to phenobarbital on its enhancement of 2-acetylaminofluorene-induced hepatic tumourgenesis in the rat, *Cancer Res.,* 33, 2701, 1973.
169. **Solt, D. and Farber, E.,** New principle for the analysis of chemical carcinogenesis, *Nature (London),* 263, 701, 1976.
170. **Pitot, H. C.,** The natural history of neoplasia — newer insights into an old problem, *Am. J. Pathol.,* 89, 402, 1977.
171. **Pitot, H. C., Barsness, L., Goldsworthy, T., and Kitagawa, T.,** Biochemical characterisation of stages of hepatocarcinogenesis after a single dose of diethylnitrosamine, *Nature (London),* 271, 456, 1978.
172. **Cayama, E., Tsuda, H., Sarma, D. S. R., and Farber, E.,** Initiation of chemical carcinogenesis requires cell proliferation, *Nature (London),* 275, 60, 1978.
173. **Gronow, M. and Thackrah, T. M.,** unpublished data.

Chapter 5

NONHISTONE PROTEINS IN GENETIC REGULATION

G. S. Stein, M. Plumb, J. L. Stein, I. R. Phillips, and E. A. Shephard

TABLE OF CONTENTS

I. INTRODUCTION

For some time it has been known that the eukaryotic genome exists in the form of a protein:DNA complex and that the chromosomal proteins play a key role in both the structural and functional properties of the genome. It has been convincingly demonstrated that four of the five principal histone proteins are responsible for the primary level of DNA "packaging", that is, the formation of nucleosomes. Additionally, several lines of evidence suggest that histones, though by themselves lacking the heterogeneity requisite for sequence specific recognition, participate in conjunction with other chromosomal macromolecules in determining the availability of genes for transcription.

Despite the intensive effort of many laboratories over the past several years, far less is definitively known about the structural or biological properties of nonhistone chromosomal proteins. Our limited understanding of these chromosomal proteins is largely attributable to their heterogeneity, the relative insolubility of many nonhistone chromosomal proteins, and the very limited availability of functional assays for chromosomal components that unequivocally reflect the situation in intact cells. However, it is apparent from present data that amongst the nonhistone chromosomal proteins are macromolecules that are involved in genome structure, in enzymatically catalyzed nuclear reactions, and possibly in determining whether or not specific genetic sequences are available for transcription.

In this chapter we will consider the involvement of nonhistone proteins in the regulation of gene expression with emphasis on a possible role for a subset of these complex and heterogeneous macromolecules at the nuclear and/or transcriptional level. We will begin by summarizing what is known about the structure and organization of several classes of eukaryotic genes since such knowledge is a prerequisite to understanding how a regulatory macromolecule might function. Experimental data which suggest a structural and/or regulatory role for nonhistone proteins will be critically evaluated, and within this context, experimental approaches which can further enhance our understanding of the properties of putative eukaryotic regulatory macromolecules will be discussed. Throughout this chapter, we will attempt to relate genomic structure and regulation because we strongly hold the conviction that the two are functionally inseparable.

II. STRUCTURE AND ORGANIZATION OF THE EUKARYOTIC GENOME

Understanding the representation and organization of DNA sequences in the genome and knowing the modes of expression of various sequences are a necessary basis for evaluating mechanisms operative in gene control and for designing experimental approaches to identify, isolate, and characterize regulatory molecules. With respect to the potential for transcription, the genomic DNA sequences of most higher eukaryotic cells can be broadly divided into those which can be transcribed and those which are not. Of those which are transcribed, the ultimate gene products can be RNAs or proteins, with examples of both types of genes being represented as unique or reiterated sequences. Globin and ovalbumin genes are examples of protein-coding DNA sequences present as single copies per haploid genome, whereas histone genes are reiterated. Most genes thus far identified which code for nontranslated RNAs such as tRNAs, 18S ribosomal RNAs, 28S ribosomal RNAs, and 5S RNAs are present in multiple copies. Whether or not such genes are expressed, and the extent to which they are transcribed, varies under different biological circumstances. Far less is known about the representation, organization, and function of nontranscribed genetic sequences although it is becoming increasingly apparent that these sequences may be involved in gene regulation and structure. For example, the putative regulatory sequences located upstream from protein coding genes are not transcribed and the AT-rich reiterated sequences which are involved in the formation of chromosomal centromeres do not undergo transcription; the latter sequences may serve as genomic sites for interaction with cytoskeletal elements that form the mitotic spindle. Hence, for a comprehensive understanding of regulation of eukaryotic gene expression, it is necessary to consider the possibility that chromosomal proteins modulate the functional and structural properties of both transcribed and nontranscribed sequences.

With the recent development of recombinant DNA technology and procedures for rapid DNA sequencing, we are beginning to appreciate that several distinctly different patterns of eukaryotic gene organization exist. For example, the genes which code for 5.8S, 18S, and 28S ribosomal RNAs are clustered, tandemly repeated, and confined to the nucleolar organizer region of the genome. These sequences are transcribed by RNA polymerase I into 45S RNA precursors which undergo a series of cleavages and chemical modifications to yield the functional ribosomal RNA species. Globin and ovalbumin genes present as single copies per haploid genome, are transcribed by RNA polymerase II, and the transcripts are extensively processed prior to translation. The protein coding sequences are interspersed with nontranslated sequences which are spliced out during post-transcriptional processing. Upstream from the mRNA coding regions of the globin and ovalbumin genes are nontranscribed sequences putatively involved in RNA polymerase binding and thus implicated in regulation of expression. Histone genes are also transcribed by RNA polymerase II, but unlike globin and ovalbumin sequences, the histone genes are reiterated. Invertebrate histone genes are organized in a series of tandem repeats, each repeat containing an H1, H2A, H2B, H3, and H4 histone coding sequence; while in mammalian cells histone genes are clustered but probably not tandemly repeated. Of potential importance with respect to the understanding of regulation is the apparent absence of nontranslated intervening sequences in each of the histone protein genes. For example, 5S ribosomal RNAs are transcribed by RNA polymerase III from reiterated genomic sequences, and it appears that expression involves sequences located within the RNA-coding region of the gene.

With our rapidly accumulating knowledge of the structure and organization of numerous eukaryotic genes, it is becoming clear that there are many different forms in which genes and gene clusters are organized. It may therefore be reasonable to anticipate that there is a comparable diversity in the properties of regulatory macromolecules and in the mechanisms by which they function.

Also important for understanding the control of gene expression is an appreciation of the mechanism by which meters of DNA are packaged as chromatin within the confines of the cell nucleus. It is necessary to establish the types of interactions which occur between structural proteins and specific genetic sequences or regions thereof. Defining alterations in such protein-DNA interactions as a function of whether or not genetic sequences are transcribed could provide a valuable insight into structural-functional features of genome regulation.

In conjunction with physicochemical studies,[1-5] early nuclease digestion studies of intact chromatin revealed that the primary level of genome packaging is in the form of a "string of beads" structure composed of nucleosomes connected by micrococcal nuclease (MNase) sensitive linker DNA.[6-8] The nucleosome is a complex of a histone octamer $(H3:H4:H2A:H2B)_2$ and 140 to 145 base pairs of DNA which are wound in 1.75 turns around the protein core in a left-handed toroidal coil.[4-13] This core particle is derived from limit MNase nuclear digests, and the 140 to 145 base pair repeat appears to be invariant throughout the eukaryotic kingdom.[6,14-16] More recent evidence suggests that the invariant repeat may in fact be equivalent to two full turns of DNA around the histone core, corresponding to 160 base pairs of DNA, which appear as a transient intermediate in limited MNase nuclear digests.[17-19] In very limited MNase nuclear digests, DNA repeats of between 160 to 240 base pairs have been identified and this repeat has been attributed to the species and tissue specificity of the size of linker DNA connecting any two core particles, and varying between 0 to 80 base pairs of DNA in length.[6,14-16] The length of linker DNA also varies as function of cellular differentiation,[20-22] a variability which may be related to the transcriptional activity of the cell.[16]

The lysine-rich histone H1 (and H5 in avian chromatin) is probably associated with the linker DNA, and may be specifically implicated in stabilizing the common "entry" and "exit" point of the two turns of DNA around the histone octamer.[17-19] Furthermore, the lysine-rich histones are capable of forming "head to tail" homopolymers *in situ*[9,10] which may cross-link nucleosomal filaments and thus provide the molecular mechanism for higher order chromatin structures.[23-28,30,31] Many structures have been proposed for higher order H1-mediated chromatin condensation, but the two most favored at present are the solenoidal model proposed by Klug and co-workers[26,27,30,31] and the superbead model proposed by Stratling and co-workers.[32-35]

As will be discussed in a later section, actively transcribed chromosomal genes have a higher sensitivity to exogenous endonucleases than the bulk chromosomal sequences. This implies that genomic transcribed sequences are maintained in a conformation which is distinguishable from bulk chromatin by virtue of a less stable nucleosomal complex and the local depletion of lysine-rich histones. The latter may be related to the specific decondensation of chromatin, thus making specific genomic domains accessible to intermolecular interactions (by nucleases and polymerases).

Table 1
CORRELATIVE EVIDENCE FOR NONHISTONE
PROTEINS AS REGULATORY MOLECULES

Ref.

Heterogeneity	52, 77, 92
Tissue and species-specificity (qualitative)	36, 37, 39, 40, 46, 52, 58, 59, 61, 62, 65, 66, 70—73, 75, 76, 78—80, 93—106, 109—112
Tissue specificity	39, 56, 59, 63, 109, 113—119
Active vs. inactive chromatin	47, 55, 57, 120—137
Changes in gene activity	
Cell proliferation	38, 41—44, 138—149
Development and differentiation	39, 40, 60, 69, 73, 78, 109, 150, 163
Gene activation by hormones and drugs	158, 164—170
Carcinogenesis and cell transformation	67, 101, 103, 148, 149, 171—189, 226
Nonhistone protein phosphorylation	
Tissue specificity	51, 52, 62, 76, 94, 112
Active vs. inactive chromatin	48, 49, 57, 244
Cell cycle (continuously dividing cells)	44, 45, 54, 68, 87, 190—193
Stimulation of cell proliferation	44, 139, 244, 194—203
Differentiation and development	69, 117, 158, 159, 161, 162, 204
Gene activation by hormones and drugs	50, 160, 205—218
Carcinogenesis and cell transformation	53, 172, 184, 219—225
Steroid hormone action	58, 63, 100, 227—241
DNA-binding nonhistone proteins	62, 64, 74, 80—86, 88—91, 105, 107, 108, 242, 243

The involvement of nonhistone proteins in this conformational difference is exemplified by the high mobility group proteins HMG14 and HMG17 whose presence is a prerequisite for the preferential sensitivity of transcribed chromatin to DNase I (see following sections). Similarly, there is some evidence of enzymatic modulation of chromatin structure by the postsynthetic modification of the DNA and histone proteins in the nucleohistone filament, activities which by definition require the presence of nonhistone proteins (histone deacetylases, histone kinases, DNA-methylases, etc.) which are directly involved in potentiating genomic domains for transcription.

III. CORRELATIVE EVIDENCE FOR NONHISTONE CHROMOSOMAL PROTEINS AS REGULATORY MOLECULES

Among the general requirements that a group of molecules may be expected to meet if they are involved in selective gene expression, are to have considerable heterogeneity, to exhibit some tissue and species specificity, to vary in their distribution between transcriptionally "active" and "inactive" chromatin, to be able to interact with specific DNA sequences and certain hormone-receptor complexes, and to undergo quantitative and qualitative changes at times of alteration of gene expression. The evidence presented in Table 1 suggests that components of nonhistone chromosomal proteins meet many of these requirements and as such, provides correlative evidence for the possible involvement of nonhistone chromosomal proteins in the control of gene readout.

A. Heterogeneity of Nonhistone Chromosomal Proteins

With present levels of resolution attainable by two-dimensional gel electrophoresis, the nonhistone proteins in chromatin have been shown to contain several hundred distinct species.[52,77,245] At this level of resolution their complexity rivals that of cytoplasmic proteins[77] and contrasts markedly with the comparative lack of heterogeneity of the histones.

B. Tissue and Species Specificity

Several groups have analyzed nonhistone proteins by one-dimensional electrophoresis and found tissue-specific differences.[36,37,39,40,46,59,62,66,70,93-98] Interpretations of these findings are complicated by the different extraction methods used by the various groups. For example, methods which extract a more representative fraction of the proteins show far less tissue specificity.[59,72] The conclusion that nonhistone proteins have only limited tissue specificity was supported by the finding that most of the major proteins were present in most tissues examined.[71,99]

Tissue specificity has also been demonstrated by immunological techniques involving, in particular, DNA-protein complexes of low molecular weight proteins with a high affinity for DNA.[58,84,100-105] This argues against the possibility that differential extractability is the main reason for some of the observed tissue-specific differences in the electrophoretic patterns of the proteins.

Wu et at.[106] compared liver and kidney nonhistone proteins from such diverse species as rat, cat, cow, chicken, turtle, and frog, and found that the protein pattern had changed a great deal during evolution. Nevertheless, each tissue contained a tissue-specific subset of proteins that appeared to have been conserved.

The introduction of two-dimensional electrophoretic techniques for the separation of nonhistone proteins resulted in better resolution than was possible using one-dimensional electrophoresis, and initial results obtained using these techniques did not change the general concept of the limited tissue- and species-specificity of these proteins.[61,73,75,76] However, with the introduction of high resolution two-dimensional electrophoretic systems,[79,245-247] it was found for example that as many as half the nonhistone chromosomal proteins of Friend and HeLa cells differed.[78] This degree of divergence was less than that observed for the cytoplasmic proteins (three quarters of which differed) indicating that as suggested by the early studies, many of the nonhistone chromosomal proteins may be under relatively strict evolutionary conservation. The tissue- and/or species-specific chromosomal proteins presumably include those involved in the control of cell-specific gene expression, whereas the remainder of the proteins may be conserved because they have to interact with DNA or other molecules in general processes, such as the maintenance of chromosomal structure, DNA replication or transcription, that are common to both cell types. One group of nonhistone proteins that has little tissue or species specificity is the HMG proteins, and these proteins have been shown to bind DNA.[107,108,250-255] Other workers have also presented evidence of the evolutionary conservation of a group of DNA-binding nonhistone proteins.[221]

A disadvantage of comparing nonhistone chromosomal protein patterns derived from whole tissues is that tissues usually consist of several different cell types, and consequently different chromatins, which complicates results concerning the tissue specificity of nonhistone proteins. Fujitani and Holoubek[109] went some way towards overcoming this problem by comparing nonhistone proteins from three different anatomical regions of brain. However, the anatomical regions still contained various nuclear types. Another approach to this problem has been to compare nonhistone chromosomal proteins extracted from fractions enriched in a particular nuclear type, for example, neuronal and glial nuclei[110-112] and more recently these proteins have been extracted from three to five types of brain nuclei prepared by isopycnic zonal centrifugation.[52,73] Although these nuclear types were not pure, this approach, which reveals cell-specific variations in nonhistone chromosomal proteins, goes a considerable way towards overcoming the problem of cross-contamination of nuclear types.

C. Quantitative Differences in Nonhistone Chromosomal Proteins of Various Tissues

Although histones are present in chromatin from most tissues at about a 1:1 (w/w) ratio to DNA, the proportion of nonhistone proteins in chromatin from different tissues varies considerably. This was first noted by Mirsky and Ris.[113] Although many tissues have a nonhistone protein to DNA ratio of about one or slightly less,[59,114] the ratio can be as low as 0.13 in sea-urchin sperm[115] or as high as 9.1 in slime mold,[116] and the amount of nonhistone protein in chromatin has been correlated with the RNA synthetic activity of that chromatin.[56,109,117] The nonhistone protein content of chromatin has been found to change with the biological state of a tissue; for example, the levels declined (along with DNA template activity) during the maturation of sperm,[118] increased during sea-urchin embryo development,[39,119] and paralleled chromatin template activity during the estrogen-induced differentiation of chick oviduct.[63]

D. Differences in Nonhistone Chromosomal Proteins of "Active" and "Inactive" Chromatin

Quantitative variations in the nonhistone chromosomal protein content of active and inactive tissues suggest that chromatin template activity is influenced by the nonhistone chromosomal proteins, and this hypothesis is supported by the finding that when chromatin was fractionated, by several different methods, into eu- and heterochromatin, the nonhistone to DNA ratio was greater in the euchromatin fraction.[55,57,120-126] Electrophoretic analysis demonstrated that there were qualitative and quantitative differences between the nonhistone proteins from the different chromatin fractions, with each fraction possessing unique species.[125,127-132] These differences have been confirmed by immunological techniques.[133] Certain specific nonhistone proteins with defined enzyme activities, such as nuclear protein kinase[134] have been found to be preferentially localized in the active fraction of chromatin. The euchromatin fractions were transcriptively active and more diffuse in structure, whereas the heterochromatin fractions were inactive and condensed. These results indicate that some nonhistone proteins may be involved in transcriptional activity and/or maintenance of a diffuse chromatin structure, whereas others may be involved in the inhibition of transcriptional activity or in the condensation of chromatin structure. However, the legitimacy of these chromatin fractions is controversial, and although there is evidence that genes that are in a transcriptionally "active" configuration are associated with the active fraction of chromatin,[123,135,136] other workers found that this was not the case.[37]

E. Changes in Nonhistone Chromosomal Proteins Associated with Modifications in Gene Activity

A regulatory role for components of the nonhistone chromosomal proteins is suggested by changes in these macromolecules that are associated with modifications of gene expression in a number of biological situations. For example, such changes in nonhistone proteins have been observed following the stimulation of resting cells to proliferate; during differentiation and development; after hormone and drug stimulation; during the transformation of cells; and with aging.

1. Changes in Nonhistone Chromosomal Proteins Associated with Cell Proliferation

When nondividing cells are stimulated to proliferate, an increase in nonhistone chromosomal protein synthesis has been observed in many different systems: salivary glands stimulated by isoproterenol;[38] lymphocytes stimulated by phytohemagglutinin,[43] concanavalin A,[139,140] leukoagglutinin,[141] or anti-immunoglobulin;[140] liver regeneration following partial hepatectomy; and refeeding of starved *Physarum poly-cephalum*[44] and *Tetrahymena*.[143-145] Electrophoretic analysis has revealed that in many cases changes in the total amount of the nonhistone chromosomal proteins are due to changes in the levels of a few specific proteins.[44,141,146-149] The increases in specific nonhistone chromosomal proteins which occur when cells are stimulated to divide precede increases in RNA and DNA synthesis and thus it is reasonable to postulate that these proteins are involved in the control of gene transcription at the onset of cell proliferation. However, the levels of some nonhistone proteins decrease when cells are stimulated to divide and these may be involved in the maintenance of the quiescent state.

2. Changes in Nonhistone Chromosomal Proteins during Development and Differentiation

During the course of development and differentiation, batteries of genes are activated and others are repressed; thus, developing and differentiating systems provide good opportunities to study the molecules involved in controlling selective gene expression. Quantitative and qualitative developmental stage-specific changes in the nonhistone chromosomal proteins have been observed in sea urchin embryos,[39,60,69,150,151] *Xenopus laevis* tadpoles,[152] *Oncopeltus* (milkweed bug) embryos,[153] and during the development of chick oviduct[63] and embryonic red blood cells.[154] These changes in the amount of nonhistone proteins paralleled changes in transcriptional activity[63,151,153,154] and structure[151] of chromatin.

Changes have also been observed in the complement of nonhistone chromosomal proteins during the differentiation of pollen from *Hippeastrum belladonna*,[155] during the conversion of lymphoid spleen to an erythroid organ,[156] and in the dimethyl sulf-oxide-stimulated erythroid differentiation of Friend cells.[78,157] Again, these changes parallel changes in the RNA synthetic capacity of the tissues. Further evidence for a link between the level of nonhistone chromosomal proteins in a cell and the RNA synthetic capacity of that cell is provided by the reduction in amount and the loss of specific nonhistone proteins observed during the condensation of chromatin and concomitant repression of RNA synthesis during spermatogenesis[158,159] and the maturation of erythroid cells.[40,160] However, in contrast to the above findings, some groups have found only minimal changes in the complement of nonhistone chromosomal proteins during prostaglandin and cAMP-induced differentiation of neuroblastoma cells,[161] during the normal or hydrocortisone-induced embryonic development of the neural retina,[162] and during the normal development of various brain tissues.[73,109,163]

3. Changes in Nonhistone Chromosomal Proteins Associated with Gene Activation by Hormones and Drugs

Several hormones activate the expression of specific genes in their target tissues, and increases in the synthesis and amount of nonhistone chromosomal proteins have been demonstrated to occur during gene activation by insulin,[164] aldosterone,[165] and phenobarbital.[166] The conclusion that these proteins are somehow involved in the hormonal control of gene expression is further supported by the evidence that the synthesis of specific nonhistone proteins is augmented at times of gene activation by cortisol,[256] and specific proteins accumulate in the "puffed" regions of insect chromosomes after treatment with ecdysone.[258] These changes are steroid and target-tissue specific. The changes in nonhistone proteins that occurred in prostate gland in response to testosterone were negated when the antiandrogen, cyproterone acetate, was administered together with testosterone in vivo.[170]

4. Changes in Nonhistone Chromosomal Proteins Associated with Carcinogenesis and Cell Transformation

Another finding which supports the hypothesis that the nonhistone chromosomal proteins are involved in the control of gene expression is that they undergo quantitative and qualitative changes during carcinogenesis and transformation, and some of these changes are evident within hours of the treatment of normal tissues with carcinogens. Although no changes were found in the complement of rat liver nonhistone proteins 3 hr after administration of diethylnitrosamine, their rate of synthesis had increased by 50% 1 hr after the administration.[171] However, when rats were treated with thioacetamide, substantial changes were found in both the levels and rates of synthesis of specific liver nonhistone proteins.[67] Similar results have been obtained for cancer of the breast,[172] liver,[148,173,174] colon,[175,176] brain,[177,178] prostate,[179] and lymphocytes.[149] However, in some cases, such as the treatment of human diploid fibroblasts with UV radiation or N-acetoxy-2-acetylaminofluorene, no changes were observed in the nonhistone proteins.[180] In the above studies differences between the nonhistone proteins of normal and neoplastic tissues were detected by one- or two-dimensional gel electrophoresis, but it has been confirmed, by immunological techniques,[103] that there are differences between the nonhistone proteins of normal and neoplastic tissues. A particularly interesting observation by Busch's group was that a protein which appeared in lymphocytes only when they became leukemic corresponded to a protein detected in hepatomas but not in normal liver.[148,149] This protein may be involved in some general mechanism of carcinogenesis such as the perturbation of replicative control.

Nonhistone chromosomal proteins also seem to be involved in mediating viral-induced modifications in gene expression. It was shown, by immunological techniques, that the nonhistone proteins of SV40-transformed WI-38 human fibroblasts differed from those of untransformed cells.[101] Gel electrophoretic and radioactive labeling studies have also demonstrated differences in the levels and rates of synthesis of specific nonhistone proteins in SV40-transformed and untransformed WI-38 cells[181-186] and in Rous sarcoma virus-infected chick embryo fibroblasts.[187] Although Gonzalez and Rees[188] did not find significant differences between the nonhistones of SV40-transformed and normal 3T3 mouse embryo fibroblasts, Krause et al.[189] found differences which were similar to those observed between SV40-transformed and normal WI-38 cells, and the latter group proposed that these changes may be a specific feature of SV40-mediated cell transformation. The possible involvement of nonhistone chromosomal proteins as mediators of aberrant gene expression in neoplastic cells has recently been reviewed.[226]

F. Phosphorylation of Nonhistone Chromosomal Proteins and the Control of Gene Expression

One aspect of nonhistone chromosomal protein metabolism that may play an important role in the control of gene expression is their rapid phosphorylation and dephosphorylation. The postsynthetic phosphorylation of proteins may well be a general method of physiological control.[259-261] Because a large proportion of the nonhistone chromosomal proteins are phosphoproteins, the evidence, discussed above, concerning the involvement of these proteins in the regulation of gene expression also applies to their phosphoprotein component. References to specific studies aimed at examining nonhistone protein phosphorylation under different biological circumstances are included in Table 1. An extensive treatment of nonhistone protein phosphorylation is in the chapter entitled "Nuclear Protein Kinases" by Mitchell and Kleinsmith in Volume III.

G. The Involvement of Nonhistone Chromosomal Proteins in Steroid Hormone Action

In addition to the findings that steroid hormones increase the rate of synthesis and phosphorylation of specific nonhistone chromosomal proteins, there is a large body of evidence which suggests that components of the nonhistone proteins mediate some of the primary actions of steroid hormones.[227-229]

After steroid hormones enter the cell they form a complex with a "receptor" protein. The hormone-receptor complex then enters the nucleus and binds to the chromatin. The nature of the specific chromatin binding sites has been extensively studied in vitro with both the isolated progesterone-receptor complex and oviduct nuclei. The hormone-receptor complex bound more extensively to chromatin than did the free hormone[230] and these workers also found that the bound hormone could be reextracted from the chromatin as a hormone-receptor complex. This result suggested that the cytoplasmic receptor is itself, at some stage, a nonhistone chromosomal protein, and is an example of the movement of some of these proteins between the cytoplasm, nucleoplasm, and chromatin. The hormone-receptor complex bound more extensively to oviduct chromatin than to chromatin of "nontarget" tissues,[230] suggesting that the chromatin of target tissue contained specific "acceptor sites" for the hormone-receptor complex.

The use of chromatin reconstitution techniques showed that the nonhistone chromosomal proteins and not the histones are involved in the tissue-specific binding of the hormone-receptor complex.[58,230] Therefore, both the receptor and acceptor molecules for progesterone may be considered nonhistone chromosomal proteins. A heterogeneous fraction of nonhistone proteins which contained the acceptor sites was isolated[63] and was shown by immunological techniques[100] to be tissue-specific. Similar results were obtained for the chromatin binding of several steroid hormones including corticosterone and estradiol,[231] thyroid hormone,[232] and testosterone.[233]

Multiple binding sites for the progesterone-receptor complex have been found,[234,235] and these binding sites have different binding affinities for the hormone-receptor complex. Most of the high affinity, tissue-specific binding sites were masked by nonhistone chromosomal proteins.[234,236] This was also found to be the case for testosterone binding sites.[233] Chromatin acceptor sites for the progesterone-receptor complex seem to be present in "nontarget" as well as "target" tissues; however, in the nontarget tissues they are completely masked by nonhistone chromosomal proteins.[229,237] In the developing oviduct, marked changes in the extent of availability of acceptor sites have been observed.[237,238] The receptor-progesterone complex seems to be associated with particular DNA sites in the genome.[229,239] However, hormone-receptor complexes bind DNA with no apparent sequence specificity[262] indicating that nonhistone chromosomal proteins may constitute an integral part of the chromosomal acceptor sites for steroid-receptor complexes, thus defining the genomic domain to be recognized and regulating the domain's availability.

More recently, the role of the steroid-receptor complexes has been called into question by the observation that steroid alone will bind specifically to target-tissue nuclear matrix preparations.[263,264] As nuclear matrix preparations represent nuclear ghosts which are depleted of histones, DNA, and nonhistone proteins,[265-268] the apparently specific ($k_m = 10^{-10}$ and 10^{-9}) binding of steroid to the residual nuclear lamina or to the interchromatinic matrix in the absence of nucleohistone and associated nonhistone proteins may require the reevaluation of current models for steroid-mediated regulation of gene expression.

It is worthy to note that nuclear matrix proteins are inevitably found as major contaminants of chromatin preparations[269] similar to those used to study chromosomal steroid receptor acceptor sites. Furthermore, DNA template processing events (transcription and replication) occur in close proximity to the nuclear lamina and nuclear pores,[267,270-274] and in the case of transcription, within access of the nuclear pores and associated filaments which may be involved in the translocation of nuclear transcripts to the cytoplasm. Similarly, it is possible that multiple forms of steroid and steroid-receptor acceptor sites coordinately regulate gene expression.

Since in their target tissues steroid hormones increase the number of sites available for in vitro transcription,[240] and activate the in vivo transcription of defined mRNA sequences such as ovalbumin mRNA,[241] the mediation of their action by nonhistone nuclear proteins associated directly with the chromosomal nucleohistone, or indirectly with nucleohistone-associated nuclear structures, is strong evidence supporting the role of these proteins in selective gene expression.

H. DNA-Binding of Nonhistone Chromosomal Proteins

Selective gene expression implies the restriction of some genetic loci and the specific activation of others, and there is evidence supporting the hypothesis that nonhistone chromosomal proteins play a role in these processes. If this is indeed the case, it is reasonable to postulate that the mechanisms through which transcriptional control is achieved will depend, in part, on associations between these proteins and DNA. In prokaryotes there are several examples of DNA-binding proteins that regulate RNA synthesis; for example, the *lac* repressor[275,276] which inhibits transcription of the *lac* operon, and the cAMP receptor protein which selectively facilitates initiation of *lac* and *gal* mRNAs.[277,278]

Specific DNA-binding nonhistone proteins have been isolated by several groups of workers. In this section we will discuss some of the properties of these DNA-binding proteins which may have a bearing on their postulated role in the control of gene expression.

1. Sequence Specific DNA-Binding Nonhistone Chromosomal Proteins

Nonhistone chromosomal DNA-binding proteins have been routinely isolated by affinity chromatography using immobilized homologous and heterologous DNA. Only a small proportion of the proteins in nonhistone preparations bind specifically to homologous DNA,[64] and many of these are phosphoproteins.[83] Considerable confusion exists as to the heterogeneity of these proteins, some groups observing a highly complex population of nonhistone proteins,[62,83] whereas others do not.[64,81,84] Similarly, most groups have found that NHPs bind selectively to homologous DNA,[62,64,82,105] but some have found little[310] species-specific binding and argue that their interaction with DNA has severely restricted the evolutionary diversity of these proteins. These data must be considered with some caution as binding conditions vary from group to group, different nonhistone protein preparations are used, in most cases considerable overlap of proteins in DNA-binding fractions has been observed, and rechromatography of "specific" DNA-binding protein fractions has yielded contradictory results.[64]

Specific binding of protein(s) to DNA probably occurs under a limited set of conditions, and optimization of specific binding in vitro requires the determination of a number of binding parameters. The nitrocellulose filter-binding of nucleoprotein complexes but not of free DNA has provided a rapid and flexible DNA-binding assay. Sevall and co-workers[90,279,280] for example, used tandem heterologous- and homologous-DNA affinity chromatography to fractionate a rat liver nonhistone protein preparation, and then used a filter-binding assay to analyze the specificity of the protein fractions. At saturation, the heterologous DNA-binding proteins bound 60 to 70% of whole rat liver DNA, whereas the homologous DNA-binding proteins bound only 35% of the total DNA, thus implying that some sequence specificity was being observed. Competition experiments established that the homologous DNA-binding protein fraction bound homologous DNA preferentially, with no such preference being observed for the heterologous DNA-binding protein fraction. Analysis of the DNA retained by the filters showed that it was enriched some fourfold in unique DNA sequences and depleted in repetitive sequences relative to whole DNA.

Cloned genomic DNA sequences have also been used to identify sequence specific binding. Hsieh and Brutlag[281] isolated a protein from *Drosophila* embryos which specifically bound a single cloned repeating unit of satellite DNA when the plasmid and its insert were in a superhelical conformation. Restriction endonuclease mapping of the nucleoprotein complex revealed that the protein protected one specific restriction site on the insert DNA fragment. Similarly, a DNA-binding protein has been isolated from *Drosophila* cells,[282] and bound to random or restriction fragments of whole cell DNA. Bound DNA was recovered from nitrocellulose filters, inserted into plasmids, and amplified by bacterial transformation. The affinity of the protein (DB-2) for cloned DNA fragments was then analyzed using a nitrocellulose filter binding assay, and bound clones reamplified by bacterial transformation. This selective process yielded a plasmid which was specifically bound by the DB-2 protein in very high ionic strengths (2.0 *M* NaCl). Immunocytochemical localization of the DB-2 protein revealed that it was concentrated specifically on chromosome 3 (section 95a/b) in intact cell nuclei, supporting the postulate that the protein is associated with a specific region of the *Drosophila* genome in vivo.

Binding of chromosomal proteins to soluble chromatin preparations has also attracted much attention recently. Dastugue and Crepin[283,284] bound a nonhistone protein fraction to homologous- or heterologous-soluble chromatin and found that binding of homologous protein to homologous chromatin stimulated transcription by exogenous RNA polymerase B compared to native chromatin preparations, and the stimulation of specific transcription by bacterial RNA polymerase was also observed. This stimulation of a functional activity was confined to homologous systems, little or no stimulation being observed in heterologous protein/chromatin mixtures. The rationale of this approach is clearly that specific binding of protein to DNA may be influenced by the conformation of the DNA in chromatin in vivo. A model system used to examine this idea was reported by Chao's group[285,286] who used a 203 base pair DNA fragment containing the *Escherichia coli Lac* operator sequence to reconstitute nucleosomal structures with calf thymus histone proteins. The nucleosomes were then bound in vitro by lac-repressor protein, and specific sequence recognition identified without the displacement of the histones. As binding was quantitative and specific, the results suggested that the reconstitution protocol yielded structures where the bound DNA strand always faced the outer surface of the nucleohistone complex. Chemical cross-linking of the histones inhibited high affinity binding of the repressor molecule, indicating that conformational changes in the nucleosome were necessary for high affinity binding. Similarly, they were able to show that high affinity binding could only occur if the operator sequence was positioned in the region of the nucleosome stabilized by H2A-H2B interaction (but not H3-H4 interactions). This implies that sequence-specific binding in eukaryotic chromatin may be modulated by DNA interactions with histone proteins which will define the accessibility of sequences to intermolecular interactions.

An alternative approach to the identification of specific DNA:nonhistone protein complexes formed in vivo has been provided by Bekhor's group.[287-291] Essentially, chromatin was extensively washed with 2.0 *M* NaCl (pH 8.0) to remove loosely bound NHPs and the histone proteins, and the protein depleted DNA dialyzed against 10 m*M* Tris-HCl (pH 8.0). Low speed centrifugation yielded a nucleoprotein complex (DNA-P) constituting some 1% of the total nuclear DNA complexed to a specific nonhistone protein fraction (the M3 fraction) which could be eluted from the DNA with 3.0 *M* NaCl, whereas binding to bacterial DNA was inhibited in 200 m*M* NaCl. The DNA in the nucleoprotein DNA-P complex was enriched in unique genomic sequences, enriched some 20-fold in transcribed DNA sequences, depleted of nontranscribed sequences and inactive gene sequences, and depleted of highly repetitive DNA sequences.

As the proteins bind specific sequences with such a high affinity, rearrangement during nuclear processing is unlikely, and the DNA-P complex probably represents a biologically significant nucleoprotein complex formed in vivo.

Another technique which may be invaluable in the identification of sequence-specific DNA-binding proteins is the protein blotting technique in which proteins are first separated electrophoretically, transferred to nitrocellulose filters, and then examined for specific binding to purified DNA sequences.[292,293] However, the specific binding of DNA sequences to immobilized proteins after transfer remains for the moment unproven.

2. Single Stranded DNA-Binding Proteins

Preferential binding of nonhistone proteins to immobilized denatured DNA has been assayed ever since it was observed that such proteins stimulated DNA synthesis in prokaryotes.[294,295] Eukaryotic single stranded DNA-binding proteins have been subsequently identified in a host of species and tissues.[74,85,91,242,243,296-300] Some of these proteins have the characteristics of DNA-unwinding proteins,[85,242,300] stimulate in vitro DNA synthesis by eukaryotic DNA polymerase A,[296,297,300] and some have nicking-closing enzymatic activities.[300] These in vitro activities may be modulated by phosphorylation,[296,301,302] but a definitive correlation between these in vitro properties and a functional role in vivo remains to be proven.

3. Other Properties of DNA-Binding Proteins

Other properties of nonhistone DNA-binding proteins have been examined in an attempt to elucidate a functional role for the proteins in vivo. For example, a protein isolated from DB cell lysates had a preferential affinity for single stranded-DNA and was a potent nuclease inhibitor in vitro.[303] DNA-binding proteins have been implicated in the cessation of gene activity,[324] inversely related to the rate of cell growth,[74,305] and were preferentially synthesized during the induction of transformed rat cells.[306] Similarly, the proteins form species and tissue specific deoxyribonucleoprotein complexes which are specific immunogens and which have been used for immunocytochemical localization studies.[84,100,307-309]

While undoubtedly DNA-binding represents a powerful tool for isolation of subsets of the chromosomal proteins, caution must be exercised in assuming that the types of protein-DNA interactions which are observed by these affinity methods are biologically meaningful insofar as they are equivalent to those that occur in the nuclei of intact cells.

IV. MORE DIRECT EVIDENCE FOR NONHISTONE CHROMOSOMAL PROTEINS AS REGULATORY MOLECULES

A. Effects of Nonhistone Chromosomal Proteins on Chromatin Template Activity

The evidence already discussed indicates that there is a correlation between the amount and type of nonhistone chromosomal proteins and the RNA synthetic capacity of a tissue. More direct evidence for the involvement of these proteins in the control of gene transcription comes from experiments in which the effects of these proteins on chromatin template activity were studied by the incorporation of radioactively labeled nucleotides into RNA in the presence of bacterial or homologous RNA polymerase. Studies carried out in the laboratories of Gilmour and Paul,[310-312] Chiu et al.,[359] and Wang and Kamiyama,[314,315] suggested that nonhistone chromosomal proteins or fractions thereof stimulated RNA synthesis from rat liver chromatin. This activation was later shown to be specific to the tissue of origin of the nonhistone proteins.[95,310-312,316,359] Results from similar studies suggested involvement of nonhistone proteins in determining variations in chromatin-template activity during the cell cycle.[317,318] Nonhistone chromosomal proteins were also found to stimulate the template activity of pure DNA using bacterial[62,319] or homologous RNA polymerase.[320] Transcription was only stimulated by homologous nonhistone proteins and was due to an increase in the number of RNA chains initiated. The protein fractions that stimulated transcription were rich in phosphoproteins and there was a correlation between their level of protein kinase activity and their degree of activation of RNA synthesis.[321-323] Dephosphorylating the nonhistone proteins abolished their stimulatory effect.[320] These results support correlations, discussed previously, between the degree of nonhistone chromosomal protein phosphorylation and the transcriptional activity of a tissue.

A loosely bound, phosphate-rich nonhistone chromosomal protein fraction which stimulates transcriptional activity, and a tightly bound nonhistone chromosomal phosphoprotein which inhibits transcription have also been isolated from Ehrlich ascites cells.[325] The inhibitory effect of the latter was not altered by the addition of excess RNA polymerase, but was reduced by increasing the amount of DNA, indicating that the inhibitory reaction involved the template and not the enzyme. The transcription stimulator bound selectively to unique sequences of DNA and activated their transcription[326] whereas the inhibitor protein bound to, and inhibited the transcription of, only reiterated DNA sequences.[327] The same group of workers isolated a nonhistone protein from calf thymus which had a similar inhibitory effect on transcription. However, the calf thymus and Ehrlich ascites inhibitors differed in molecular weight and subunit composition, and only inhibited transcription from homologous DNA.[464] Thus, both the stimulatory and inhibitory proteins act in a tissue-specific manner which suggests that they are capable of recognizing and binding to specific DNA sequences. Several other groups have also isolated nonhistone chromosomal proteins which stimulate eukaryotic RNA polymerase II[328-332] or I.[330,333,334] In one case, a protein factor which enhances the formation of initiation complexes[332] was found to be active only when another "helper" protein was bound to the polymerase.[335]

While these results are indeed consistent with a regulatory role for components of the complex and heterogeneous nonhistone chromosomal proteins, caution must be exercised in interpretation of data from these experiments. Direct addition of nonhistone chromosomal proteins to protein-DNA complexes may result in nonspecific aggregation. Additionally, results from such experiments do not provide definitive answers with respect to regulation of specific genetic sequences.

B. Nonhistone Chromosomal Proteins and Transcription of Specific Genetic Sequences

Despite a number of limitations, the techniques of chromatin reconstitution, in vitro transcription and the availability of complementary DNA probes for specific genetic sequences have to some extent facilitated assessment of the involvement of nonhistone chromosomal proteins in the transcription of several genes; the globin, ovalbumin, and histone genes have been studied most extensively. The regulation of globin and ovalbumin gene expression are reviewed by others.[336-340] In this chapter we will focus on studies carried out in this laboratory directed towards understanding the role of nonhistone chromosomal proteins in the transcription of histone genes during the cell cycle.

1. Evidence for Cell Cycle Stage-Specific Histone Gene Transcription

In agreement with the well-documented S phase specific synthesis of histone (reviewed in Reference 313 and seen in Figure 1), several lines of evidence suggest that transcription of histone mRNA sequences is restricted to the S phase of the cell cycle. Early evidence which was interpreted to support synthesis of histone mRNA only dur-

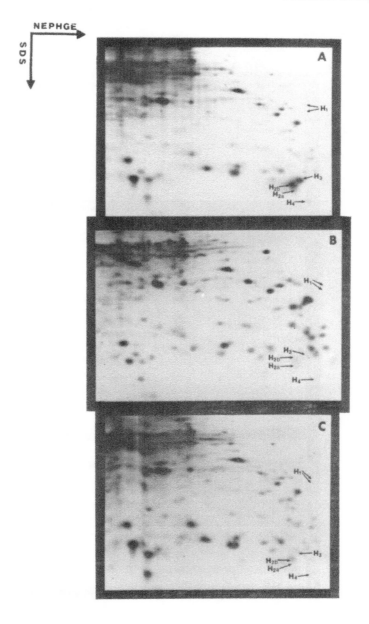

FIGURE 1A. Two-dimensional electrophoretic analysis of ³⁵S-methionine-labeled total cellular proteins.
(A) S phase, (B) G1 phase, (C) Cytosine arabinoside-treated S phase. HeLa cells in suspension culture were
synchronized by double thymidine block (2 mM). After release from the second thymidine block, cells were
maintained at a density of $5 \times 10^5/1$ ml for 1 hr. S phase cells were incubated for an additional 2 hr at 37°
either in the presence or absence of the DNA synthesis inhibitor, cytosine arabinoside (40 μg/ml). Ten
milliliters of either S phase or cytosine arabinodide-treated S phase cells were centrifuged, washed in spinner
salts solution containing 5 μM methionine (37°C), and resuspended in 2.8 ml of low (5 μM) methionine
medium (containing cytosine arabinoside when appropriate). ³⁵S-methionine (140 μCi) was added and cells
were pulse labeled for 45 min at 37°. G1 phase HeLa cells were obtained by mitotic selective detachment
and labeled as described above for S phase cells. Cells were harvested by centrifugation at 1500 × g for 5
min, rinsed with ether, and dried before solubilizing in 9.5 M urea-2% (w/v) NP-40-2% (w/v) Ampholine-
5% (w/v) mercaptoethanol-0.3 M NaCl-1.0 mg/ml protamine sulfate. ³⁵S-labeled peptides were fractionated
in a two-dimensional electrophoretic system in which the first dimension was nonequilibrium pH gradient
electrophoresis (NEPHGE) and the second dimension was in SDS-containing, 15% (w/v) acrylamide gel.

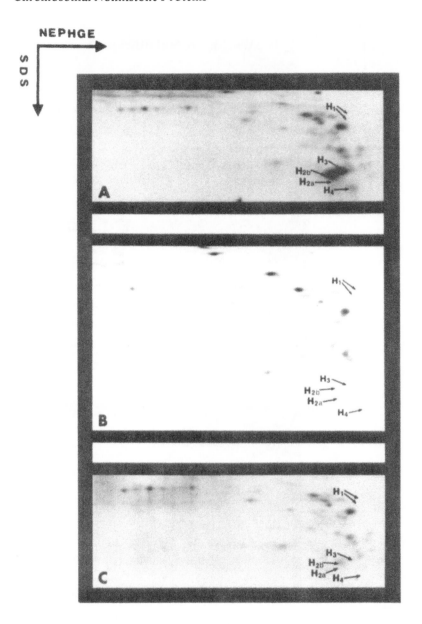

FIGURE 1B.   Two-dimensional NEPHGE/SDS electrophoresis of acid-extracted nuclear proteins of (A) S phase, (B) G1, and (C) Cytosine Arabinoside-treated S phase cells. ^{35}S-methionine-labeled cells were lysed in 10 mℓ KCl-10 mM Tris-1.3 mM MgCl$_2$ (pH 7.4) containing 0.65% (v/v) Triton® X-100 and nuclei were pelleted by centrifugation at 800 × g. Nuclear pellets were extracted with 0.4 M H$_2$SO$_4$ for 30 min and the acid soluble nuclear proteins were precipitated from the supernatant at −20° overnight after addition of 3 volumes of 95% ethanol. Ethanol-precipitated proteins were rinsed with ether and dried before solubilizing in 10 M urea-0.2% SDS — 30 mM lysine — 2.5 mM ZnCl$_2$ (pH 4.8). Protein concentrations were adjusted to 2 mg/mℓ and NaCl was added to a final concentration of 0.4 M. Samples were digested with S1 nuclease (Sigma). Two-dimensional electrophoretic analysis of ^{35}S-labeled peptides was performed as described in the legend of Figure 1A.

ing S phase in continuously dividing HeLa cells came from pulse labeling experiments in which ^{3}H-uridine-labeled 7-9S RNAs were isolated from the polysomes of S phase cells but not from the polysomes of G$_1$ cells.[341] This observation is consistent with the possibility of transcriptional control of gene expression, an interpretation which is further supported by in vitro translation[342,343] and nucleic acid hybridization,[344,345] data suggesting the presence of histone mRNA sequences on the polysomes, in the postpolysomal cytoplasmic fraction, and in the nuclei of S phase but not G_1 phase HeLa cells. Other data suggesting that control of histone gene expression may be mediated transcriptionally come from experiments which show that nuclei[345,348] and chromatin[345,346] of S phase, but not G_1 HeLa cells, transcribe RNAs which hybridize with a ^{3}H-labeled DNA complementary to the mRNAs for the H2A, H2B, H3, and H4 histones or with a ^{3}H-labeled H4 cDNA. Similar results were obtained following stimulation of nondividing human diploid fibroblasts to proliferate.[349]

While taken together these results seem to support control of histone gene readout residing at least in part at the transcriptional level, caution must be exercised in interpreting the in vitro translation and transcription experiments. The limits of detection of histones synthesized in vitro when very small quantities of RNA template are available leave something to be desired. Interpretation of data from in vitro nuclear transcription experiments is complicatd by an inability to establish definitively the initiation of RNA chains in vitro and by the presence of endogenous nuclear histone mRNA sequences. Results from in vitro chromatin transcription experiments can be misleading since transcription was carried out with bacterial RNA polymerase, and endogenous chromatin-associated RNA sequences can also interfere with evaluation of nucleic acid hybridization. It must also be pointed out that because, in the experiments just described, hybridization was carried out in RNA excess using unlabeled RNA and a ^{3}H-labeled histone DNA complementary to the mRNAs for the H2A, H2B, H3, and H4 histones, the presence of rapidly turning over histone mRNA sequences in the nuclear RNA or in the nuclear or chromatin transcripts of G_1 cells might not have been detected. In fact, Melli et al.[350] have detected the presence of histone mRNA sequences in the nucleus throughout the cell cycle of HeLa cells by hybridizing ^{3}H-labeled RNA with excess sea urchin histone DNA. However, there are some reservations concerning the experiments of Melli et al.[350]: (1) The method utilized for cell synchronization was double thymidine block, a procedure which results in at least 20% of the G_1 and G_2 cells actually being S phase cells (as measured by ^{3}H-thymidine labeling and autoradiography). This is in contrast to experiments in which the HeLa cDNA probe was used and in which G_1 cells were obtained by mitotic selective detachment, a procedure which yields a population of G_1 cells containing less than 0.1% S phase cells. (2) Hybridization analysis of HeLa cell RNAs was carried out with sea urchin DNA and there appears to be only limited sequence homology between human and sea urchin histone sequences.

Determination of the presence or absence of histone mRNA sequences in G_2 HeLa cells is complex. By nucleic acid hybridization analysis we have observed a limited representation of histone mRNA sequences in the polysomal, postpolysomal cytoplasmic, and nuclear RNA fraction of G_2 cells.[351] The rate of hybridization of these subcellular RNA fractions from G_2 cells with histone cDNA is approximately 20% of that observed with similar RNA fractions of S phase cells. However, thymidine labeling followed by autoradiography indicates that at least 20% of the G_2 population (obtained by double thymidine block or mitotic selective detachment) consists of cells which are undergoing DNA replication.[352] Because effective methodologies are not available for obtaining a pure population of G_2 phase HeLa cells, it is difficult to determine unequivocally whether or not G_2 cells contain histone mRNA sequences.

2. Chromosomal Proteins and Regulation of Histone Gene Expression

If control of histone gene expression is mediated at the transcriptional level, it becomes important to identify and characterize the nuclear and chromatin components (chromosomal proteins and/or RNAs) which render histone sequences transcribable.

If control of histone gene expression is at a posttranscriptional level and the synthesis of histone mRNA sequences occurs throughout the cell cycle, it is still important to determine the mechanism by which the genes are rendered transcribable and maintained in a transcriptionally active structural configuration. Posttranscriptional regulation of histone gene expression during the cell cycle also raises a broad spectrum of questions relevant to posttranscriptional mechanisms: Are there variations in the rate of turnover of histone mRNA sequences in various intracellular compartments during the cell cycle? Are there cell cycle stage-specific changes in the abilities of histone mRNA sequences to be translated in vitro? Are there differences in proteins associated with histone mRNA sequences during the cell cycle? Are there variations in the presence or activity of enzymes involved in the processing and/or degradation of histone mRNA sequences during the cell cycle? Are there changes in the binding of histone mRNAs to ribosomes during the cell cycle, resulting from alterations in histone mRNAs and/or the properties of the ribosomes? Unfortunately, to date, little is understood regarding processes associated with posttranscriptional regulation of the expression of protein-coding genes in eukaryotes and most of these questions are beyond the scope of this chapter.

a. Nonhistone Chromosomal Proteins and Histone Gene Transcription in Continuously Dividing Cells

A role for nonhistone chromosomal proteins in the regulation of gene expression during the cell cycle has been suggested by several lines of evidence. Variations observed in the composition and metabolism of the nonhistone chromosomal proteins during G_1, G_2, S, and mitosis and their correlation with changes in transcription are consistent with a regulatory function for these proteins (reviewed in References 227, 353). Further evidence that nonhistone chromosomal proteins may be responsible for specific transcription at various stages of the cell cycle comes from a series of chromatin reconstitution studies which indicate that nonhistone chromosomal proteins are involved in determining the quantitative differences in availability of DNA as template for RNA synthesis during the cell cycle of continuously dividing cell,[354] as well as after stimulation of nondividing cells to proliferate.[355]

To examine the involvement of nonhistone chromosomal proteins in rendering histone genes available for transcription in vitro from chromatin during the cell cycle, we have pursued the following approach. Chromatin isolated from G_1 and S phase cells was dissociated in 3 M NaCl, 5 M urea, and each chromatin preparation was fractionated into DNA, histones, and nonhistone chromosomal proteins. Chromatin preparations were then reconstituted by the gradient dialysis procedure of Bekhor et al.[356] utilizing DNA and histone pooled from G_1 and S phase cell and either G_1 or S phase nonhistone chromosomal proteins. Essentially DNA, histones, and nonhistone chromosomal proteins were combined in high salt-urea and the salt was progressively removed by stepwise dialysis, followed by removal of the urea. In vitro RNA transcripts from chromatin reconstituted with G_1 nonhistone chromosomal proteins and from chromatin reconstituted with S phase nonhistone chromosomal proteins were annealed with histone cDNA. RNA transcripts from chromatin reconstituted with S phase nonhistone chromosomal proteins hybridized with histone cDNA, while those from chromatin reconstituted with G_1 nonhistone chromosomal proteins did not exhibit a significant degree of hybrid formation. It should be emphasized that the kinetics of hybridization with the cDNA were the same for transcripts of native S phase chromatin and transcripts of chromatin reconstituted with S phase nonhistone chromosomal protein (Cr_0t $1/2 = 2 \times 10^{-1}$). Furthermore, the amount of RNA transcribed and the recoveries during isolation of these transcripts from native and reconstituted preparations were essentially identical. These results[346] suggest the possibility that nonhistone chromosomal proteins may be involved in determining the availability of histone sequences for transcription from chromatin during the cell cycle. Such a regulatory role for nonhistone chromosomal proteins is in agreement with results from several laboratories which have pointed to these proteins as potentially important for tissue-specific transcription of globin genes[357-359] and ovalbumin genes.[336]

However, extreme caution must also be exercised with regard to in vitro chromatin transcription and interpretation of results from such experiments. Meeting the following criteria should be considered minimal to justify chromatin as a bona fide in vitro system which reflects with complete in vivo fidelity the transcription of histone mRNA sequences.

1. The probe used to identify histone mRNA sequences should be homologous and should accurately reflect the complete mRNA. Eventually it will be desirable for the probe to represent all the nucleotide sequences of the primary gene transcript. Specific probes should be available so that hybridization analysis of histone gene transcripts can be carried out in RNA as well as in DNA excess.
2. Conditions for RNA transcription, isolation of transcripts, and hybridization analysis should be such that endogenous, chromatin-associated RNAs do not interfere with quantitation of histone gene transcripts.
3. Precautions must be taken to eliminate or account for RNA-dependent RNA synthesis when transcription is carried out with *E. coli* RNA polymerase. Eventually it will be desirable to transcribe chromatin with the appropriate homologous RNA polymerase — in which case it would be necessary to demonstrate that only the enzyme responsible for transcription of histone genes in vivo is operative in the in vitro system.
4. It should be established that initiation of RNA synthesis is taking place in vitro, not merely elongation of RNA molecules initiated in vivo.
5. It would be desirable to establish that initiation and termination of transcription in vitro from chromatin are at the sites on the DNA molecule at which they occur in vivo.
6. It should be shown that the degree of symmetric or asymmetric transcription of histone genes from chromatin is the same as that which takes place in intact cells.
7. In addition to establishing fidelity of histone gene transcription from chromatin in vitro it is necessary to show that the availability *and* lack of availability of other genetic sequences for in vitro transcription from chromatin is the same as that which takes place in intact cells.
8. The inhibitory effects of histones on DNA transcription in vitro require that studies, particularly after reconstitution, are performed with chromatin preparations whose conformation reflects that found in vivo.

Not all of the above criteria need to be met for in vitro chromatin transcription to be a useful model system for examining regulation of histone gene expression. In fact, failure to meet certain of these criteria may provide information regarding structural and functional properties of chromatin. However, in order to consider in vitro chromatin transcription justifiable for studying the regulation of histone gene expression, it must be unambiguously shown that histone mRNA sequences are transcribed in vitro only from chromatin derived from cells in which the genes are transcribed in vivo.

b. Regulation of Histone Gene Expression in Human Diploid Cells Stimulated to Proliferate

The control of histone gene expression was similarly studied in nondividing human diploid fibroblasts following stimulation to proliferate.[349] Hybridization analysis of

RNAs isolated from various intracellular compartments suggests the presence of histone mRNAs only in cells undergoing DNA synthesis, and transcription of histone mRNA sequences only from chromatin of S phase cells. Chromatin reconstitution experiments provide evidence for a possible role for the nonhistone chromosomal proteins in the regulation of histone gene transcription.

3. Utilization of Cloned Human Histone Genes for Studying Histone Gene Expression

From the preceding discussion it is apparent that further assessment of the regulation of human histone gene expression, and particularly the involvement of chromosomal proteins in control of these genetic sequences, would be significantly facilitated by the availability of cloned homologous histone genes. Therefore in this section we will briefly discuss the structure and organization of human histone genes and the manner in which clones containing these sequences can be utilized to isolate and characterize molecules potentially involved in regulation of the structure and transcription of histone genes.

a. Structure and Organization of Human Histone Genes

We isolated a series of 12 genomic clones containing histone coding sequences, their flanking sequences, and noncoding sequences from a lambda charon 4A gene library constructed by Maniatis and collaborators.[29] These clones were analyzed by hybridization with heterologous probes as well as by hybrid selection-translation, and restriction enzyme mapping. While the human histone genes are clustered, unlike what has been observed in sea urchin and *Drosophila,* human histone genes are not arranged in the form of a unique tandem repeat. Rather, human histone genes exhibited at least three types of arrangements with respect to restriction sites and the order of coding sequences; each of these arrangements is clearly distinguishable from the others. *In situ* hybridization studies[360,361] and restriction analysis data from human-rodent hybrids[362] suggests that in humans the histone genes may be clustered on the distal end of the long arm of chromosome seven in the G negative band Q—34. Partial restriction maps of each of the three types of arrangements of human histone genes, representing information from seven individual genomic clones, are shown in Figure 2.

Of particular interest with respect to the structure, organization, and expression of human histone genes is an observation we made several years ago of multiple forms of human H4 and H3 histone mRNAs.[363-365] In our initial studies we demonstrated that two different mRNA species coding for H4 histones can be isolated from polysomes of S phase HeLa cells. The two variants were separated by polyacrylamide gel electrophoresis of 5-18S polysomal RNA extracted from synchronized cells. Upon elution from the gel both species were assayed for template activity in vitro using the wheat germ cell-free protein synthesizing system. Both RNAs were translated into H4 histone. Using several types of denaturing gels (98% formamide-acrylamide, 10 mM glyoxal-acrylamide, and methyl mercury-agarose), the two mRNAs were found to retain their distinct mobilities, indicating the two are different in molecular weight. Characterization of the 3'-terminus of each RNA by oligo dT-cellulose column chromatography revealed no significant poly(A) region on either molecule. After treatment of the RNAs with T1, T2, and pancreatic ribonucleases followed by electrophoresis on DEAE paper, it was shown that both species had a capped 5'-terminus.[248] Sequence analysis was conducted by digesting the two RNA species with T1 ribonuclease and subsequently performing a two-dimensional fractionation of the resulting oligonucleotides. Although the majority of oligonucleotides from the two H4 mRNAs migrated similarly, a few variations were noted, suggesting some sequence heterogeneity. This is similar to Grunstein's findings for the C1 and C3 fractions of H4 histone mRNA from sea urchin.[366] Electrophoresis of the in vitro translation products of the HeLa

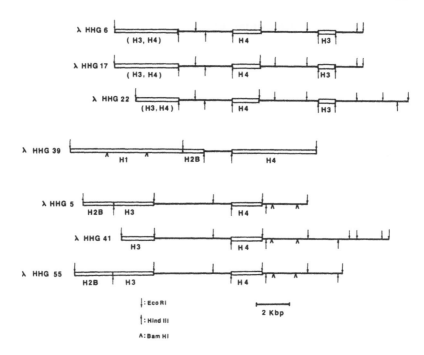

FIGURE 2.   Restriction maps of the seven λHHG recombinant phage. Eco RI, Hind III, and Bam HI restriction sites were determined, and coding regions were assigned by hybridization with specific histone gene probes. For purposes of comparison the insert of λHHG 22 is displayed in an orientation (with respect to λ) opposite to that of the other six λHHG clones. Boxes indicate restriction fragments to which histone-coding sequences have been assigned.

histone H4 mRNAs in the acetic acid-urea system could not separate the independently synthesized products which co-migrate with marker H4. One-dimensional tryptic peptide mapping of the translation products labeled with several different radioactive amino acids revealed no differences between the two molecules. Two-dimensional tryptic peptide mapping has also been carried out and again no differences were detected.[363-365]

More recently using higher resolution fractionation procedures, we observed at least seven different H4 histone mRNA species associated with the polysomes of S phase HeLa S3 cells.

1.   Electrophoretic fractionation of in vivo synthesized, ^{32}P-labeled S phase HeLa cell histone mRNAs under denaturing conditions (6.0% (w/v) acrylamide — 8.3 M urea run at 50°C) reveals three major H4 histone mRNAs (Figure 3).
2.   Subsequent fractionation of the three H4 histone mRNAs under nondenaturing conditions, where advantage was taken of possible variations in secondary structure, resulted in separation of each band into several additional H4 histone mRNA species (Figure 4). Each of these mRNA bands was excised from the gel, heat denatured in urea, and refractionated electrophoretically in nondenaturing and denaturing gels. Each of the individual H4 mRNA bands reran as a single species under both nondenaturing (Figure 5) and denaturing conditions, indicating that these RNA bands are not artifacts of the electrophoresis system and suggesting that each is in fact a unique H4 histone mRNA species.

DENATURING GEL

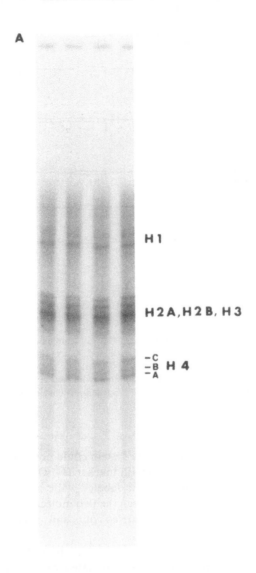

FIGURE 3.   S phase HeLa cells were ^{32}P-labeled in vivo and polysomal RNA was isolated in the presence of vanadyl ribonucleoside.[367] The RNA was fractionated on a sucrose gradient and the region between 5S and 18S was pooled. The RNA was ethanol precipitated and then dissolved in 8.3 *M* urea — 5 M*M* EDTA (pH 8.0), heated to 100° and quick cooled before loading on a 6% (w/v) polyacrylamide — 8.3 *M* urea gel buffered with 50 m*M* Tris-borate — 1 m*M* EDTA (TBE).[345] The gel was run at 20 W which gave a surface temperature of 50 to 60°C. These conditions are sufficient to denature most RNA molecules.[368] After autoradiography, the bands labeled A, B, and C were cut out and the RNA was eluted electrophoretically into dialysis tubing.

3.   Each of the numbered bands in Figure 4 (1—10) was translated in a wheat germ cell-free protein synthesizing system and each coded for only H4 histone.
4.   Each of the numbered bands in Figure 4 was also subjected to T1 ribonuclease digestion and two-dimensional fingerprinting of the resulting oligonucleotides. Variation in the oligonucleotide maps of bands 1,2,5,7,8,9, and 10 (some of which are shown in Figure 6) show sequence differences in at least seven H4

NON-DENATURING GEL

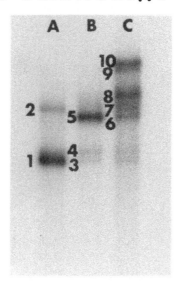

FIGURE 4.   RNAs from the gel shown in Figure 3 were ethanol precipitated, dried, and resuspended in 4 *M* urea — 2.5 m*M* EDTA (pH 8.0). The RNAs were then heated to 100° and loaded into a 6% (w/v) polyacrylamide — TBE gel with no urea[345] for fractionation under nondenaturing conditions. The numbered bands were excised and the RNAs eluted electrophoretically.

    histone mRNA species. The fingerprints also indicate that the oligonucleotides of the smaller molecular weight H4 histone mRNAs are not a subset of those generated from the larger RNAs, ruling out a simple precursor-product relationship.

5.   The identity of the H4 histone mRNA species was confirmed by hybrid selection and subsequent electrophoretic fractionation under denaturing conditions of in vivo-labeled S phase HeLa cell polysomal RNA (Figure 7).

6.   Variations were observed in the representation of H4 histone mRNA species when polysomal RNAs from HeLa cells and WI-38 human diploid fibroblasts were fractionated electrophoretically, immobilized covalently on diazotized cellulose, and identified by hybridization to ^{32}P-labeled cloned H4 histone DNA sequences (Figure 8). At least one H4 histone mRNA species that is observed in WI-38 cells is not represented in the polysomal RNA of HeLa cells. A possible difference in the expression of H4 histone mRNA species in normal and transformed cells is implied.

7.   We have begun to assign H4 histone mRNA species to individual H4 histone genes using a modification of the Berk and Sharp procedure.[249] As shown in Figure 9, when H4 histone mRNA species collectively were annealed with cloned genomic H4 histone sequences, each H4 histone gene formed an S1 nuclease resistant hybrid with only one H4 mRNA.

    We have recently initiated studies directed towards identification, isolation, and characterization of multiple forms of human H3 histone mRNAs. The approaches being pursued are similar to those we have been using to examine H4 histone mRNA species. We have been able to identify at least four distinct H3 histone mRNA species associated with the polysomes of HeLa S3 cells and WI-38 human diploid fibroblasts. These H3 mRNAs can be identified by hybridization of cloned human genomic H3 histone DNA sequences (^{32}P-labeled) to electrophoretically fractionated HeLa cell and

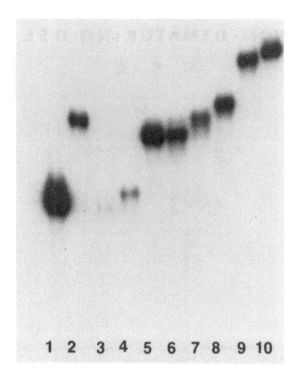

FIGURE 5.   RNAs eluted from the numbered gel bands of Figure 4 were denatured by boiling in 4 *M* urea (as described in the legend to Figure 4) and run in the same gel system. Each RNA reran as a single band with the same relative migration exhibited in the first gel (Figure 4), eliminating the possibility that the multiple bands in Figure 4 are due to conformational isomers of a single RNA species.

WI-38 polysomal RNAs covalently bound to diazotized cellulose. As shown in Figure 3, four HeLa cell H3 histone mRNAs can be isolated from total cytoplasmic RNAs by hybrid selection with cloned H3 histone sequences. Additional characterization of the H3 histone mRNAs and assignment to H3 histone mRNA species to individual H3 histone genes is being pursued.

b. Level of Control of Human Histone Gene Expression

To gain a more definitive insight into regulation of histone gene expression during the cell cycle, we have used cloned genomic human histone sequences to reexamine the representation of histone mRNAs in the nucleus and cytoplasm of G1 and S phase synchronized HeLa S3 cells. In agreement with earlier observations from our laboratory[344-346,348,349,364,365,373] and others where histone mRNAs were analyzed by RNA excess hybridization with homologous histone cDNAs, histone mRNA sequences are present in significant amounts in the nucleus and cytoplasm of HeLa cells only during S phase when histone protein synthesis occurs.

The representation of HeLa cell histone mRNAs was assayed by hybridization to cloned H4, H3, H1, and H2B histone sequences. To standardize conditions for the hybridization, we took advantage of the long-standing observation that the inhibition of DNA replication results in a rapid loss of histone mRNAs from the polysomes of S phase HeLa cells. As shown in Figure 10 under the hybridization conditions employed in our studies, ^{32}P-labeled plasmid DNAs containing histone sequences anneal with polysomal RNAs from S phase HeLa cells fractionated electrophoretically in methylmercury agarose gels and transferred electrophoretically to diazotized cellulose. Consistent with the anticipated loss of histone mRNA sequences from polysomes following

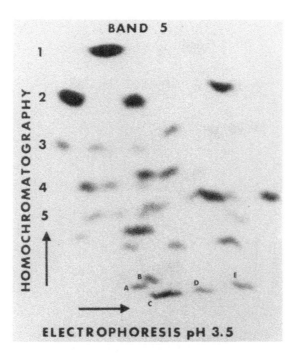

FIGURE 6A. RNAs eluted from the gel shown in Figure 4 were digested with Tl ribonuclease and the resulting oligonucleotides were fractionated on cellulose acetate at pH 3.5 in 7 *M* urea in the first dimension followed by homochromatography on PEI-cellulose thin layer plates using homomixture B of Volkaert and Fiers[369] in the second dimension. The band numbers correspond to those in Figure 4. Bands 5 and 6 (Figures 6A and B) are identical, as is shown by the correspondence between the lettered oligonucleotide spots. Bands 1 and 10 (Figures 6C and D) are different, as shown by spots a, b, and c in 1 which are not found in 10, and spots d and 3 from 10 which are not found in 1. Also note that both 1 and 10 differ from 5 and 6.

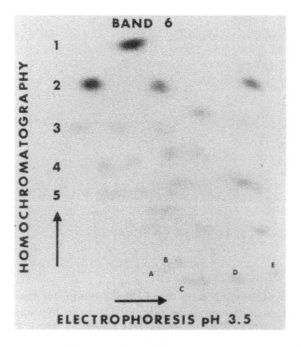

FIGURE 6B.

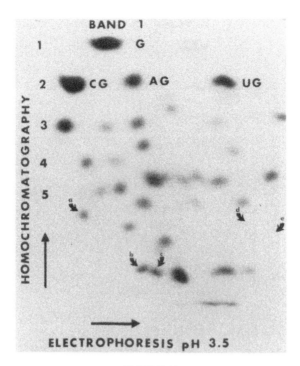

FIGURE 6C.

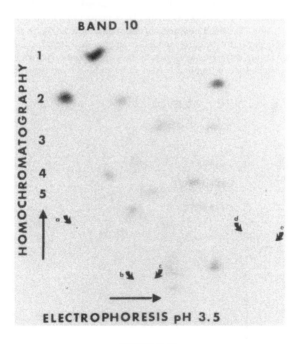

FIGURE 6D.

inhibition of DNA synthesis, a greater than 95% inhibition was observed in the abilities of the radiolabeled, histone-containing plasmid DNAs to hybridize with filtered-immobilized polysomal RNAs from S phase HeLa cells in which DNA synthesis was blocked to a comparable extent by treatment with cytosine arabinoside. The sensitivity of our hybridization assay is such that we can detect less than 500 pg of histone mRNA.

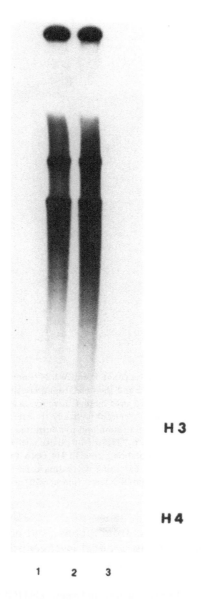

H 3

H 4

1   2   3

FIGURE 7.   Fifty micrograms of plasmid DNA (containing H3 and H4 human histone genes) were linearized with Eco RI, treated with proteinase K in 2% (w/v) SDS, and extracted with phenol-chloroform-isoamyl alcohol (25:24:1). The DNA was denatured with 0.5 M NaOH, neutralized with NaCl-Tris-HCl and passed through 13 mm Sartorius® nitrocellulose filters. The filters were baked at 80° in vacuo for 2 hr and hybridized with ^{32}P-labeled total cytoplasmic RNA from S phase HeLa cells in 1.5 ml Eppendorf® tubes in 50% (w/v) formamide — 0.5 M NaCl — 10 mM Hepes (pH 7.3) — 1 mM EDTA — 0.2% (w/v) SDS at 47° for 40 hr. The filters were washed 10 times with 1 ml of 1 × SSC — 0.5% (w/v) SDS at 68° and 3 times with 1 ml of 2 mM EDTA (pH 7.0) at 68°. The hybridized RNA was eluted with two 300 μl aliquots of distilled water at 100° for 2 min, ethanol precipitated, and run on an 8.3 M urea denaturing gel (as in Figure 1) (Lane 3) along with total cytoplasmic RNA (Lane 1) and RNA from hybridization mix which did not bind to filter (Lane 2).

To address the level at which regulation of histone gene expression occurs during the cell cycle in HeLa S3 cells, we assayed the abilities of ^{32}P-labeled (nick-translated) histone DNAs to hybridize with filtered-immobilized RNA from G1 and S phase cells. The rationale was that because histone synthesis is confined to S phase hybridization

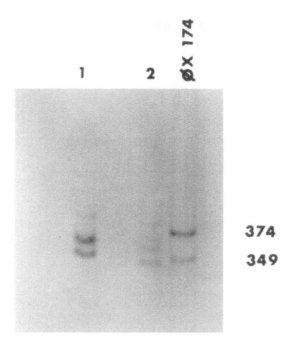

FIGURE 8.   Seventy-five micrograms of HeLa (Lane 1) and WI-38 (Lane 2) cytoplasmic RNAs were preincubated in 50 mM methylmercury hydroxide and then electrophoretically fractionated on 3% (w/v) low gelling temperature agarose gels containing 5 mM methyl mercury. After a neutralization/equilibration step,[370] the nucleic acids within the gel were electrophoretically transferred to diazotized cellulose as described by Stellwag and Dahlberg.[371] Prehybridization and hybridization solutions were as described[370] except that 5× Denhardt solution (minus BSA), 0.7 mg/ml of carrier nucleic acid and 0.1% (w/v) SDS were used. The paper was hybridized with a radiolabeled, human H4 DNA fragment (1 × 10⁷ cpm/μg, 1 × 10⁶ cpm/ml) at 50° for 36 hr. Washing was at 68° using decreasing concentrations of SSC. The paper was blotted dry and autoradiographed using Kodak® XAR-5 film at −70° with a Cronex® "Lightning Plus" intensifying screen.

of the histone DNA probes to RNAs from S phase, but not from G1 cells, would be consistent with nuclear and/or transcriptional level control. However, hybridization to both G1 and S phase RNAs would suggest that regulation of histone gene expression resides at a posttranscriptional step.

As shown in Figure 11, hybridization between λHHG55 DNA, a recombinant Charon 4A phage containing human genomic H2B, H3, and H4 histone sequences and G1 nuclear or cytoplasmic RNAs was barely detectable, while hybridization of λHHG55 DNA with both nuclear and cytoplasmic RNAs of S phase HeLa cells was apparent. In these experiments G1 cells were obtained by mitotic selective detachment, a procedure which yields a G1 population containing less than 0.5% S phase cells. Both G1 and S phase RNA were fractionated electrophoretically with 100/μg. Ethidium bromide staining indicated that similar amounts of 18S and 28S RNAs from G1 and S phase cells were fractionated in the gels and transferred to diazotized cellulose (greater than 90% transfer was obtained) in both cases. Because there is an increase in the amount of ribosomal RNA per cell in S phase compared with G1 the hybridization observed is probably an underestimation of the amount of histone mRNA in an S phase cell. On longer exposure of the "northern blot" some hybridization of the H3-H4-H2B probe with G1 RNA becomes apparent (approximately 0.1% the level observed with S phase RNA); this amount of annealing is within the limits that can be expected from the number of S phase cells in a G1 population — the biological limits of the system.

A   DENATURING   GEL

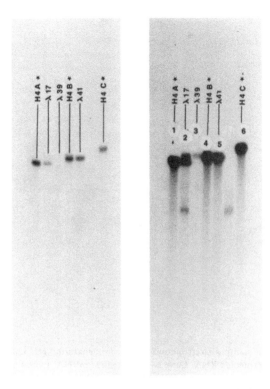

FIGURE 9A.   Identification of H4 histone mRNA species which are S1 nuclease resistant after hybridization to histone DNA clones. 10 to 20 μg of λHHG recombinant DNAs were mixed with 500 to 2,000 cpm of ^{32}P-labeled total H4 histone mRNA isolated from a gel similar to the one shown in Figure 3 and precipitated with ethanol in a 1.5 ml microfuge tube. The pellet was drained thoroughly, and dissolved in 20 μl redistilled but not deionized formamide (BRL). Four lambdas of 2 M NaCl, 0.4 M Pipes (pH 6.4), 5 mM EDTA was added and the tube was placed at 80°C for 5 min. Hybridization was at 50°C for 5 hr. After hybridization, 225 λ of ice cold 0.25 M NaCl, 0.025 M Na acetate (pH 4.4), 0.00045 M ZnSO$_4$ was added and the tube was mixed thoroughly and immediately placed on ice. One hundred ten units of S1 nuclease (Sigma) in 5 λ were then added, and the sample incubated at 37°C for 30 min. Twenty lambdas 10% (w/v) SDS, 20 λ 0.2 M EDTA (pH 8.0), 20 μg tRNA, 200 λ of phenol, and 200 λ chloroform/isoamyl alcohol (24:1 v/v) were added, and the mixture was vortexed, centrifuged, and the supernatant was ethanol precipitated. The dried pellet was dissolved in 8.3 M urea, 5 mM EDTA, heated to 100°C for 2 min and electrophoresed on an acrylamide gel. S1 resistant samples electrophoresed in an 8.3 M urea gel. Lane 1—H4A histone mRNA. Lane 2—λHHG 17 protected RNA. Lane 3—λHHG 39 protected RNA. Lane 4—H4B histone mRNA. Lane 5—λHHG 41 protected RNA. Lane 6—H4C histone mRNA. Left panel shows light exposure and right panel shows dark exposure.

When the northern blot of G1 and S phase HeLa cell RNAs shown in Figure 11 was rehybridized with ^{32}P-labeled plasmid DNA containing different H3 and H4 histone coding sequences, S phase specific hybridization was also observed (Figure 12). Similarly, S phase specific hybridization was observed when these northern blots were hybridized with a plasmid containing H2B and H1 histone sequences (Figure 12).

An unexpected result (Figure 13) was obtained when a northern blot of polysomal RNA from G1 and S phase HeLa cells was hybridized with ^{32}P-labeled DNA from λHHG41A, a recombinant Charon 4A phage containing H3 and H4 coding sequences derived from another human histone gene unit. While hybridization with S phase but not G1 histone mRNAs was observed, intense hybridization was also seen with an RNA of approximately 330 nucleotides present predominately in G1 cells.

B NON-DENATURING GEL

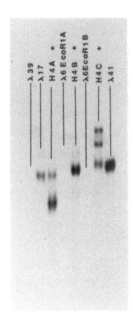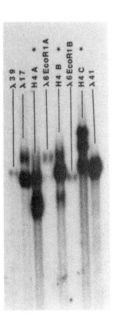

FIGURE 9B.   S1 resistant samples electrophoresed in a nondenaturing gel. Lane 1—λHHG 39 protected
RNA. Lane 2—λHHG 17 protected RNA. Lane 3—H4A histone mRNA. Lane 4—λHHG 6 Eco RI fragment
A protected RNA. Lane 5—H4B histone mRNA. Lane 6—λHHG 6 Eco RI fragment B protected RNA.
Lane 7—H4C histone mRNA. Lane 8—λHHG 41 protected RNA. Left panel shows light exposure and right
panel shows dark exposure.

The increase in the representation of both nuclear and cytoplasmic histone mRNAs
in S phase compared with G1 cells suggests that histone genes are preferentially expressed during a restricted period of the cell cycle when DNA replication occurs. Since
synthesis of histone proteins is also confined to S phase, it is reasonable to postulate
that nuclear and/or transcriptional level control is operative. Transcriptional regulation of histone gene expression during the cell cycle of HeLa S3 cells is consistent with
earlier observations that histone cDNA hybridized preferentially with in vitro-synthe-
sized RNAs (nuclear and cytoplasmic) and in vitro chromatin transcripts of S phase,
but not G1 cells.

However, an unequivocal demonstration that regulation of histone gene expression
during the HeLa cell cycle resides solely at the transcriptional level requires establishing
that (1) the very limited presence of histone mRNA sequences in G1 cells obtained by
mitotic selective detachment is attributable to those few S phase cells in the G1 population (approximately 0.5%) and (2) histone gene transcription is initiated only during
S phase.

Although Melli and co-workers[350] reported the presence of approximately equivalent
amounts of histone gene transcripts in G1 and S phase HeLa cells, these results are
not surprising since histone sequences were assayed by hybridization under nonstrin-
gent conditions with a heterologous probe — sea-urchin histone DNA sequences cloned
in lambda. Also in the experiments of Melli and co-workers,[350] cells were synchronized
by double thymidine block which yields a G1 population containing more than 20% S
phase cells.

We cannot completely dismiss the possibility that regulation of histone gene expression during the HeLa cell cycle may to some extent be mediated posttranscriptionally.

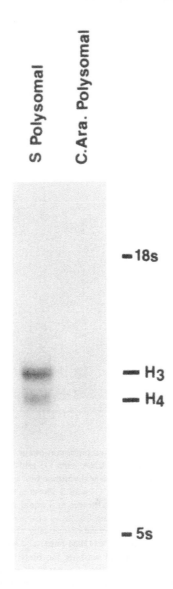

FIGURE 10.   Hybridization of electrophoretically fractionated polysomal RNA from S phase and cytosine arabinoside-treated S phase HeLa cells with ³²P-labeled human genomic H3 and H4 histone sequences. 100 µg of total HeLa cell polysomal RNAs were electrophoretically fractionated in a 2% (w/v) agarose gel containing 5 m*M* methylmercury hydroxide, electrophoretically transferred to DBM paper, and hybridized with ³²P-labeled (nick-translated) plasmid DNA containing human genomic H3 and H4 histone sequences.

However, the low steady state level of histone mRNA sequences we observed in the nuclei of G1 cells would indicate that if histone sequences are transcribed at the same rate throughout the cell cycle they must be rapidly degraded in G1 cells. Alternatively, it is possible that histone sequences are transcribed at a much lower level outside of S phase. Such low level transcription might be functional or reflect "leaky" transcription of some or all of the histone genes. Within this context the persistence of a limited amount of histone transcripts from the previous S phase in the subsequent G1 should be considered.

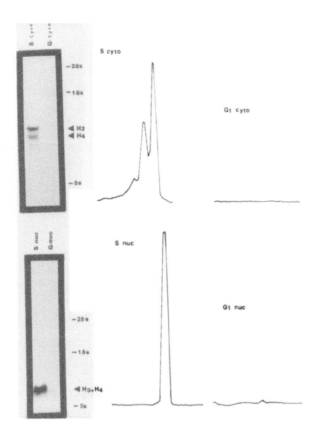

FIGURE 11.   Hybridization of human genomic recombinant phage λHHG 55 DNA containing H3 and H4 histone sequences with cytoplasmic and nuclear RNA from G1 and S phase HeLa cells. S phase cells were obtained 1 hr after release from two cycles of 2 mM thymidine block while the G1 population was obtained by mitotic selective detachment. 100 μg of total G1 and S phase cytoplasmic RNA were fractionated in a 2% (w/v) agarose-5 mM methylmercury hydroxide gel and equal amounts (100 μg) of G1 and S phase nuclear RNA were fractionated on a gel containing 1% (w/v) agarose-5 mM methylmercury hydroxide. The positions of 28S, 18S, and 5S markers were determined optically after staining the gel with ethidium bromide. RNAs were electrophoretically transferred to DBM paper and hybridized with ³²P-labeled (nick-translated) λHHG 55 DNA. After hybridization and washing the blots were analyzed autoradiographically at −70° using preflashed XAR-5 film (Kodak®) with a Cronex® "Lightning Plus" intensifying screen. Autoradiograms were scanned with a Joyce-Loebel densitometer and quantitated by planimeric analysis. (A) Cytoplasmic RNAs, (B) Nuclear RNAs.

A definitive explanation for the observed hybridization of ³²P-labeled HHG41 DNA (which contains H2B, H3, and H4 human histone genes) with a 330 nucleotide polysomal RNA species present predominately in G1 cells is not yet available. However, since the G1 polysomal RNA species is encoded in a genomic sequence in a close proximity to histone coding sequences, a possible regulatory role for the G1 RNA should not be dismissed. Further analysis of the RNA and the genomic DNA sequences in which it is encoded as well as interactions of chromosomal proteins with these sequences is underway.

c. Identification, Isolation, and Characterization of Regulatory Molecules

The availability of cloned genomic human histone sequences is permitting us to gain insight into the structure and organization of human histone genes and in particular into the properties of putative regulatory sequences. Such sequences can be used for the identification and isolation of those components of the chromosomal proteins

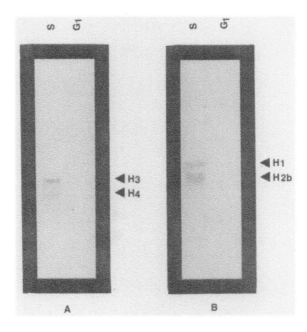

FIGURE 12. Autoradiographic analysis of G1 and S phase cytoplasmic RNAs hybridized with cloned genomic human H3, H4, H2B, and H1 histone sequences. Fifty micrograms of G1 and S phase HeLa cell cytoplasmic RNAs were fractionated electrophoretically in a 2% (w/v) agarose-5 mM methylmercury hydroxide gel, transferred electrophoretically to DBM paper, and hybridized with: (3A) ^{32}P-labeled (nick-translated) pHu 2.6H plasmid DNA containing H3 and H4 histone sequences or (3B) ^{32}P-labeled pHu 4.8E plasmid DNA containing H2B and H1 histone sequences.

which interact with DNA in a species-specific manner. Additionally, with the availability of in vitro transcription systems which initiate with in vivo fidelity, it is reasonable to be optimistic that during the next few years we will be able to significantly enhance our understanding of the manner in which histones and other eukaryotic genes are regulated.

V. NONHISTONE CHROMOSOMAL PROTEINS AND CHROMATIN STRUCTURE

A. The Nuclease Sensitivity of Transcribed Chromatin

The preferential accessibility of transcribed chromosomal DNA sequences to exogenous endonucleases, pioneered some 5 years ago by Weintraub and Groudine,[375] has been extensively used to probe the conformation of transcribed chromatin and its associated nonhistone proteins. Three principal nucleases have been used in such studies, and selected reports will be briefly reviewed below.

1. Deoxyribonuclease I (DNase I)

The sensitivity of transcribed compared with nontranscribed chromosomal DNA sequences in limited nuclear digests with DNase I has been established for a variety of species and tissues, ranging from avian erythroid globin genes[375] to integrated adenovirus genes.[376] One possible exception may be found in yeast chromatin which is uniformly digested by DNase I,[377] but as such a large proportion (40%) of the yeast genome is transcribed, it is possible that it exists in a conformation which is not directly comparable to that in higher eukaryotes.

The DNase I sensitivity of transcribed genes is independent of their rate of transcription,[376,378-381] and the sensitivity includes "quiescent genes" which are transcribed at

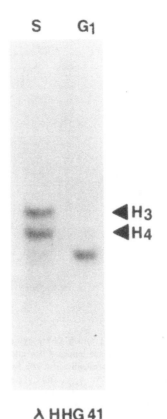

λ HHG 41

FIGURE 13.   Autoradiographic analysis of G1 and S phase RNA hybridization with
DNA from λHHG 41, a recombinant phage containing H3 and H4 human histone
sequences. The DBM blot shown in Figure 11A was incubated in sterile water at 100°
to remove ^{32}P-labeled λHHG 55 probe. After confirming that the λ55 probe was no
longer detachable, the filters containing fractionated cytoplasmic RNAs were annealed
with ^{32}P-labeled λHHG 41 DNA.

rates which are negligible compared to their fully active state (leaky genes), but which
may be induced[380-385] or which are repressed[382,383,386] as a consequence of hormonal
action, drug administration, or as a function of cellular differentiation. Thus, at an
early stage of differentiation, genomic domains are committed to a DNase I sensitive
conformation irrespective of subsequent induction or repression of transcription.

This nuclease accessibility of specific genes has now been mapped with cloned genomic DNA fragments containing both structural and flanking DNA sequences. Of
particular interest has been the observation that the 5′-terminal regions of a number
of genes are hypersensitive to DNase I.[384,386,391,392] This hypersensitivity extends some
300 base pairs upstream from structural sequences, and as these regions may contain
the putative promoter/regulatory sequences, it implies that they may be maintained in
an especially accessible conformation.

Similarly, the mapping techniques have been used to determine the sizes of DNase I
sensitive domains. The chicken β-globin gene domain for example, contains moderately DNase I sensitive DNA extending 7 to 8 kbp on either side of the transcribed
structural sequences.[386] The chicken α^d- and α^a-globin genes are present in a highly
DNase I sensitive subdomain (5 kbp) which includes 500 bp on the 5′-terminal of the
α^d-globin gene, the nontranscribed DNA between the α^a- and the α^d-genes, and extends
some 2 kb beyond the 3′-terminus of the α^a-gene. This subdomain is encompassed in

a larger domain (50 to 100 kbp) which is moderately DNase I sensitive, and which includes the embryonic chicken μ-globin gene. The μ-globin gene loses its high DNase I sensitivity and hypersensitive 5′-terminus 5 to 14 days after fertilization, a time lapse which parallels the repression of its transcription in favor of the adult α-globin genes.[382,390]

2. Deoxyribonuclease II (DNase II)

Gottesfeld and co-workers[129,395] found that chromatin solubilized by nuclear DNase II digestion could be subfractionated by precipitation with $MgCl_2$. The soluble chromatin was enriched in single copy genomic DNA sequences complementary to cellular poly (A+) mRNA sequences, it was enriched in nonhistone proteins, enriched in both nascent RNA and heterogeneous nuclear ribonucleoprotein complexes (hnRNP), and was depleted of histone H1.

The reproducibility of the fractionation protocol is somewhat variable however, as conflicting data have been reported. Lau et al.,[396] for example, repeated the experiments described by Gottesfeld and Partington[395] using dimethylsulfoxide-induced and uninduced Friend erythroleukemia cells. Whereas the latter observed a sixfold enrichment of globin DNA sequences in Mg^{++}-soluble chromatin derived only from induced cells, the former were only able to detect a 7- to 14-fold enrichment of nascent RNA in the soluble fraction, including high levels of globin gene transcripts after cellular DMSO induction.

Similarly, where specific fractionation of sequences has been observed[397,398] and despite a relative enrichment in the soluble fraction, recoveries of the sequences have been in the order of 20% of the total, the remainder being distributed in the nuclease resistant and Mg^{++}-insoluble chromatin fractions.

3. Micrococcal Nuclease Digestions (MNase)

Although it is not clear whether transcription complexes are found associated with nucleosomal DNA or not, micrococcal nuclease digests of whole nuclei have revealed that transcribed sequences are packaged as nucleosomal repeats.[381,388,390,393,399-403] More detailed analysis has revealed an inverse relationship between the relative enrichment of transcribed sequences and the size of the MNase DNA digest fragment,[399-402,404] implying that the linker DNA of transcribed chromatin is especially accessible to endonuclease activity compared with bulk nontranscribed chromosomal sequences.

Levy and co-workers[379,399,404] established that the chromatin solubilized by the lysis of MNase-digested trout testes nuclei with 1 mM EDTA could be subfractionated by the precipitation of histone H1-associated chromatin with 0.1 M NaCl as originally described by Olins et al.[405] Furthermore, the saline-soluble chromatin was highly enriched in histone H1-depleted mononucleosomes containing 140 to 200 bp of DNA which were in turn enriched some six- to sevenfold in sequences complementary to cellular poly (A+) mRNA. Thus, transcribed chromatin is believed to be decondensed by virtue of a local depletion of lysine-rich histones making the linker DNA preferentially sensitive to endonuclease digestion.

Unlike the DNase I sensitivity of genomic sequences however, transcriptionally quiescent genes are not preferentially cleaved by MNase[381,401,402] indicating that at least two different structures contribute to the nuclease sensitivity of transcribed chromatin domains. One, encompassing quiescent genes, resides at the level of the nucleosome core particle and one is retained exclusively by transcribed sequences residing at the level of the linker DNA and histone H1-associated higher order chromosomal structures. This is supported by the observation that the preferential sensitivity of transcribed sequences is still apparent in crude mononucleosome preparations.[387-390]

High resolution restriction maps of specific genes and gene clusters have been used to determine the relative sensitivity of transcribed and adjacent nontranscribed chromosomal DNA sequences. The MNase cleavage sites of the *Drosophila* 5S gene cluster, known to be arranged in a 160 to 200 tandem gene repeat, was recently mapped,[403] and found to occur in multiples of two nucleosomes implying that the nucleosomes were specifically phased. Restriction mapping of the dinucleosomal DNA revealed that the intergenic spacer DNA could exist in either of two phases: associated with the nucleosome core particle, or as part of the linker DNA between nucleosomes. It may be argued that the intergenic spacer DNA contains regulatory/promoter sequences, and these are only accessible in the phase where they are part of the internucleosomal linker DNA. On the other hand, Samal et al.,[393] also found that in the *Drosophila* histone gene cluster, the nontranscribed DNA between structural sequences was organized in one well-defined nucleosomal arrangement, whereas the nucleosomal phasing of the structural genes was quasirandom. Thus, either nucleosomal phasing is restricted to a subset of genomic sequences, or the dinucleosomal repeat is produced by some other structural phenomenon which may or may not be related to the control of transcription.

Two lines of evidence suggest that the dinucleosomal repeat in chromatin is inherent in chromatin structure as opposed to being related to the accessibility of regulatory sequences. Firstly, limited DNase I digestion of intact chromatin yields a dinucleosomal repeat[406,407] apparently due to the alternation of nuclease sensitive sites on the chromatin fiber. Secondly, electric dichroism studies of the chromatin filament[31] have suggested that the solenoidal higher structure of chromatin proposed by Klug and co-workers[26,30] may have to incorporate alternating linker DNA and associated histone H1 on the inside and outside of the solenoidal cylinder. Thus, a dinucleosomal repeat may be an inherent geometry of the chromatin fiber, defined by the alternating presence and absence of histone H1 on the endonuclease accessible surface of the solenoid.

B. The Molecular Basis for the Nuclease Sensitivity of Transcribed Domains

The nuclease digestion studies imply that transcribed genomic domains exist in a decondensed/H1-depleted conformation, and that the nucleosomal DNA of structural and flanking sequences is destabilized in favor of a high or moderate DNase I sensitivity. Three main factors are believed to contribute to this specific conformation.

1. Histone Acetylation

Sodium butyrate, a noncompetitive inhibitor of nuclear histone deacetylase[408] has been routinely used to prepare chromatin containing hyperacetylated histone H3 and H4 molecules.[408-428] Hyperacetylated histones have a reduced affinity for DNA, and their inhibitory effect on in vitro transcription is also reduced as a consequence of acetylation.[410-413] Similarly, increased levels of in vivo histone acetylation have been correlated with increased levels of in vivo RNA synthesis.[414-416]

Hyperacetylated chromatin is preferentially sensitive to DNase I digestion,[417-419] but not to MNase cleavage,[419-421] indicating that histone acetylation sensitizes chromatin to nuclease digestion at the level of the nucleosome core particle. It has been established that the acetylation of H3 and H4 molecules destabilizes the nucleosomal DNA and alters interhistone interactions.[421-423] Thus, histone acetylation may provide an enzymatic basis for modulating the nucleosomal conformation in order to facilitate gene transcription.

The use of sodium butyrate is becoming increasingly controversial as some of its effects may not be solely due to the inhibition of histone deacetylase. For example, butyrate blocks differentiation of chick myoblasts,[424] reduces specific binding of thyroid hormone-receptor complexes to rat pituitary cell nuclei,[425] induces specific gene expression in Friend erythroleukemia cells,[426,427] blocks the induction of egg white genes by steroid hormones in chick oviduct,[428] and is a potent inhibitor of DNA replication.[423] Therefore, there is some uncertainty about whether histone acetylation is the cause or effect of gene activation and, about whether the gross metabolic and morphological changes observed in cells after butyrate administration are exclusively attributable to histone deactylase inhibition.

2. DNA Methylation

Evidence implicating DNA methylation and the repression of transcription in eukaryotes has been recently reviewed by Lindahl.[429]

Detailed analysis of the methylation of specific transcribed genomic domains and their flanking sequences has received considerable attention recently. Weintraub and co-workers,[382,383,390] for example, examined the DNase I sensitivity and degree of methylation in the chick erythrocyte α-globin gene cluster as a function of gene transcription. The α^a- and α^d-globin subdomain, which includes 1 to 2 kb of flanking sequences and the intergenic spacer DNA, was highly DNase I sensitive and was undermethylated compared to the flanking nontranscribed/moderately DNase I sensitive DNA sequences. Similarly, the embryonic μ-globin gene within the α-globin gene cluster is repressed during development, an event which has been directly correlated with its methylation, its loss of DNase I hypersensitive sites, and the change from being highly DNase I sensitive in its active state, to being moderately sensitive in the quiescent state.

As well as providing a possible enzymatic modulation of gene expression, DNA methylation might also define genomic sequences in a manner that will survive one round of DNA replication.

3. The High Mobility Group Proteins (HMGs)

The HMG proteins represent the most extensively characterized nonhistone and nonenzymic proteins in the nucleus. They are readily identified and isolated due to their high solubility in mineral acids,[430-432] little tissue or species specificity,[433-435] their relatively high concentration in the nucleus (about 10^6 molecules/cell),[430,431,433-436] and their high electrophoretic mobilities in acid-urea polyacrylamide gels.[430-436] Four main HMG proteins have been identified in calf thymus (HMG1, HMG2, HMG14, and HMG17) in order of increasing electrophoretic mobility in acid-urea, and similar proteins have been identified in most other tissues and species examined.[433-435] One possible exception is trout testes which only has two major HMG proteins: The H6 protein which has some 50% sequence homology with calf HMG17,[432] and trout testes specific HMGT which has been compared to calf HMG1 and HMG2.[437]

The HMG proteins are completely eluted from chromatin with 0.35 M NaCl (pH 7.0 to 7.4).[430,433] All the HMG proteins bind DNA in vitro in low ionic strength,[250-255] and although there is no apparent sequence specificity, HMG1 and HMG2 have a preferential affinity for single stranded DNA and the potential to unwind double stranded DNA.[250-252,254,255] All the HMG proteins have also been found associated with isolated mononucleosomes derived from limited nuclear digestions with MNase.[379,399,404,437,438,445]

The possible implication of the HMG proteins in the structure of transcribed chromatin originally stemmed from reports that they were preferentially released from nuclei during limited nuclear digestions with DNase I,[437,445-448] although this is somewhat controversial.[449] The controversy arises to a certain extent because HMG protein release was not always adequately quantitated during digestion, and in certain cases, it is clearly not possible to compare the properties of HMG proteins from widely differing species. For example, preferential release of HMG1 and HMG2 has been documented for calf thymus, mouse brain, pig thymus, and Pekin duck.[439,445,446] Preferential release of H6 has also been documented in trout testes nuclear digests,[447,437,463] but H6 is sometimes compared to calf HMG17 which is not released during nuclear digests[439,445,449] in some laboratories, but is in others.[450] Similarly, HMGT which is sometimes compared to calf HMG1 and HMG2 is not preferentially released during trout testes nuclear digests.[437,462]

The HMG proteins have been found to be enriched in the DNase II-sensitive/Mg^{++}-soluble chromatin prepared by the Gottesfeld procedure.[437,451] The preferential release of HMG proteins has also been observed during limited MNase nuclear digests.[399,404,437,463] More recently however, this release has been quantitated for trout testes where only 9.0% of the total nuclear HMGT was released during MNase digests,[452] a release which approximates to that observed during MNase digestion of pig thymus nuclei.[445] On the other hand, the extractability of HMGT from nuclei is considerably enhanced after MNase digestion,[452] implying that HMGT is associated with MNase-sensitive chromatin. HMG proteins are enriched in the chromatin solubilized after nuclear digestion with MNase and nuclear lysis in 1 mM EDTA (pH 7.4),[445,452] but there is some evidence that there are at least two populations of HMG proteins in the nucleus of trout testes[447,452] and the nucleus of pig thymus;[445] one population appears to be associated with nuclease-sensitive/EDTA-soluble chromatin, whereas the second population remains associated with the nuclease resistant/EDTA-insoluble residual chromatin pellet. All the HMG proteins have been found associated with 0.1 M NaCl-soluble mononucleosomes[399,404,441,445] which are reportedly enriched in transcribed DNA sequences.[399,404] The HMGs are also present in stoichiometric amounts in 0.1 M ($Na^+ + K^+$)Cl^-/Mg^{++}-soluble mononucleosomes,[442,443] and have been found as subnucleosome complexes released during the preferential digestion of Ehrlich ascites tumor cell chromatin.[253,453]

Trout testes H6 protein has been found on core particle mononucleosomes,[399,404] implying an association with core particles in transcribed chromatin[399,404] whereas HMGT has been implicated with the linker DNA of histone H1-depleted transcribed chromatin.[379,399,404,417,437,447,463] Similarly, HMG14 and HMG17 have been found associated with core particle nucleosomes from rabbit,[438] although this was subsequently altered to an association with mononucleosomes containing a minimum of 160 base pairs of DNA.[441] However, the polar nature of the HMG14 and HMG17 molecules[252,432,454,455] is believed to permit their binding to both the histone octamer and the nucleosomal DNA in core particles, whereas HMG1 and HMG2 bind DNA, are apparently found as homopolymers in vivo,[456] and can partake in hydrophobic interaction,[457] all of which may point to their presence on linker DNA where they may either cross-link the chromatin filament by head to tail interactions with other HMG proteins or with histone H1. Alternatively, it has been proposed that they replace histone H1 on the linker DNA of transcribed chromatin[445,447,463] and are thereby involved in the decondensation of MNase-sensitive-transcribed chromatin.

The role of HMG14 and HMG17 in the structure of transcribed chromatin is less controversial. It was observed that transcribed genomic sequences in chromatin lost their preferential sensitivity to DNase I in the absence of HMG14 and HMG17, but their sensitivity could be restored by reconstitution of HMG14 and HMG17 back on to the intact chromatin.[383,386,389,390,450,458] This observation is also true for crude mononucleosome preparations,[389,390] indicating that the HMG proteins sensitize the nucleosomal DNA to DNase I in a linker DNA-independent fashion. This is supported by the observation that in whole chromatin, HMG14 and HMG17 sensitize transcribed sequences in the absence of histone H1 and H5.[389] The reconstitution studies,[389] and the use of affinity chromatography with immobilized HMG14 and HMG17[390] have established that HMG14 and HMG17 recognize 1 in 20 nucleosomes, and bind them specifically to restore the preferential DNase I sensitivity. In competition experiments, it was found that the HMG proteins bind preferentially to HMG-depleted mononucleosomes, but will preferentially bind to naked DNA rather than native nucleosome.[389] Similarly, H2a-H2b dimers compete for the nucleosomal HMG binding sites whereas H4-H3 dimers will not inhibit preferential binding, indicating that the binding sites are composed of an H3-H4:DNA complex. Mapping of the HMG-dependent DNase I sensitivity has revealed that only transcribed sequences are HMG-related, whereas flanking sequences, which are only moderately sensitive to DNase I, are not associated with an HMG-dependent sensitivity. Furthermore, HMG14 and HMG17 preferentially bind to highly transcribed genomic sequences which are undermethylated, but not to the methylated nontranscribed or quiescent genes in the genomic domains.

Thus, HMG14 and HMG17 appear to specifically bind chromatin containing specific sequences despite the fact that they bind DNA with no apparent specificity. Attempts to identify the binding site have included the use of mononucleosomes which have been washed with 0.6 *M* NaCl, or which have been dissociated in high salt and urea and then reconstituted, but in both cases specificity of binding was restored after reconstitution.[390] Furthermore, no one nonhistone protein is present on the bound nucleosomes in sufficient amounts to imply a role in HMG-nucleosome recognition. In fact, the 0.6 *M* NaCl washed mononucleosomes have no detectable nonhistone proteins.[390] Alternatively, the competition studies have implicated histone H3 and H4 histones in the nucleosome recognition, and this probably includes undermethylated DNA sequences.[382,383,386,389,390] Although the recognized nucleosomes contain hyperacetylated core histones, the levels of acetylation are below those required for the molar specificity of binding.[390]

Thus, for the present, it appears that at least three distinct structures contribute to the sensitivity of transcribed chromatin. The structure which defines bulk chromatin is probably mediated by histone H1 (and H5) and the related highly condensed higher order chromosomal structures. Transcribed chromatin is essentially depleted of the lysine-rich histones, and a high sensitivity to MNase implies an extended conformation. Genomic domains have an HMG-independent moderate sensitivity to DNase I, the domain including nontranscribed flanking sequences and quiescent genes. Thirdly, transcribed chromosomal sequences are undermethylated, have hypersensitive 5′-terminal sequences, are highly sensitive to DNase I, contain acetylated histones, and are specifically associated with HMG14 and HMG17. The two nonhistone proteins appear to be able to recognize a specific nucleosomal conformation which is probably induced by a combination of both histone and DNA postsynthetic modifications, implying that enzymic (nonhistone) activities which are present in the nucleus in substoichiometric amounts may be responsible for defining the structure of genomic chromosomal sequences to be transcribed and/or repressed.

VI. NONHISTONE CHROMOSOMAL PROTEINS AS REGULATORY MOLECULES

In this chapter we have attempted to review progress which has been made during the past several years toward elucidating the role of chromosomal proteins in determining the structural and functional properties of the genome. However, we are only at the threshold of understanding the nature of eukaryotic regulatory macromolecules and their mode of action. With regard to the regulation of gene expression, histones appear to act as nonspecific repressors of DNA-dependent RNA synthesis. In contrast, amongst the complex and heterogeneous nonhistone chromosomal proteins are components which appear to regulate the transcription of specific genetic sequences. Yet, many perplexing problems concerning the nonhistone chromosomal proteins remain to be resolved.

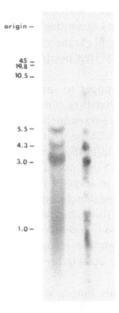

FIGURE 14A.   Hybridization of Eco RI digested λ FS 89 DNA with ³²P-labeled RNA
from S phase control and cytosine arabinoside-treated HeLa cells. DNA from clone λ
FS 89 was restricted with Eco RI, transferred to nitrocellulose, and hybridized with
³²P-labeled S phase polysomal RNA (control). Following autoradiography the hybridized RNA was eluted by heating the filters in boiling water and the filters were cut
vertically. Half of the filter was rehybridized with ³²P-labeled S phase polysomal RNAs
from cytosine arabinoside-treated cells, half was hybridized with ³²P-labeled polysomal
RNAs from cytosine arabinoside-treated cells.

The specific nonhistone chromosomal proteins responsible for rendering particular
genetic sequences transcribable must be identified. This is an especially difficult problem since to date the nature of the regulatory proteins is an enigma. It is generally
assumed that specific regulatory proteins comprise only a very small percentage of the
nonhistone chromosomal proteins. However, experimental evidence to support this
assumption is lacking. It is also reasonable to construct a viable model for gene regulation in which multiple copies of regulatory proteins are associated with the genome,
with only a limited number of these proteins existing in a "functional interaction".
An important concept which should be considered is that a single protein may regulate
several genes — particularly in situations where cellular events are functionally inter-
related or coupled. For example, one may envision several genes involved with genome
replication being controlled by a single regulatory protein. This concept is supported
by our recent identification of cloned genomic human sequences whose mRNAs, like
histone mRNAs, are present on the polysomes during S phase and are selectively lost
from the polysomes when DNA synthesis is inhibited (see Figure 14). This implies the
coordinate expression of functionally related genes is being observed. A similar situation may exist with hormone-stimulated processes. It is not clear whether regulatory
proteins should comprise a subset of the nonhistone chromosomal proteins with common characteristics such as molecular weight, charge, or structure. In this regard it

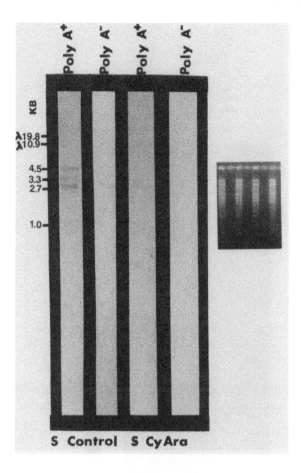

FIGURE 14B. Hybridization of poly A+ and poly A− RNAs from S phase or cytosine arabinoside-treated S phase cells to Southern blots of Eco RI restricted λ FS 89 DNA. RNAs were partially hydrolyzed by mild alkaline treatment and ³²P-labeled in vitro using T4 poly-nucleotide kinase in the presence of [δ³²P]-ATP. Note the preferential hybridization of S phase poly A+ RNA. The ethidium bromide stained gel used for the Southern blot is shown in the right side of the figure.

will be interesting to establish whether similar types of proteins control "single copy genes" such as globin genes, as opposed to "reiterated genes" such as histone and ribosomal genes.

A basic question to be answered is whether activation of genes is brought about by newly synthesized nonhistone chromosomal proteins or by modifications of preexisting genome-associated nonhistone chromosomal proteins. Alternatively, proteins residing in the cytoplasm or in the nucleoplasm may be modified in such a manner that they become associated with the genome and thereby render genes transcribable. Johnson and co-workers[139] have observed that activation of lymphocytes by mitogenic agents results in accumulation of preexisting cytoplasmic proteins in the nucleus. A nucleoplasmic pool of nonhistone chromosomal proteins has also been reported.[459] These latter two observations are consistent with the possibility that alterations in gene readout involve recruitment of proteins from the cytoplasm or nucleoplasm and their subsequent association with the genome. However, other studies suggest that protein synthesis is required for activation of transcription in human diploid fibroblasts following stimulation to proliferate.[460] In addition to numerous correlations between posttranslational modifications of nonhistone chromosomal proteins and changes in gene read-

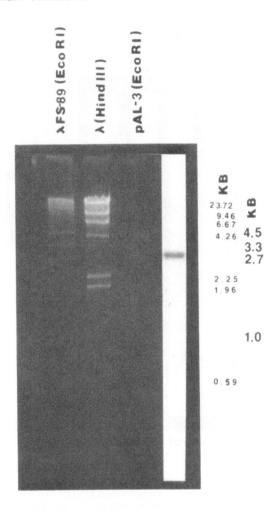

FIGURE 14C.   Identification of PAL-3 subclone of λ FS 89. PAL-3 was restricted with Eco RI and fractionated electrophoretically in a 0.8% agarose gel. Note the presence of the 4.36 Kb pBR322 band and the 2.7 Kb Eco RI fragment of λ FS 89. This result was confirmed by hybridization of nick-translated PAL-3 DNA to Southern blot of Eco RI digested λFS 89 (shown in the right lane). The bold numbers indicate the size of the Eco RI fragments of λ FS 89 and the smaller numbers indicate the size of the λ Hind III markers.

out, recent studies suggest that the phosphate groups on nonhistone chromosomal proteins are important in rendering histone genes transcribable in vitro.[461,462]

Elucidating the manner in which nonhistone chromosomal proteins are associated with other genome components should significantly enhance our understanding of the mechanisms by which regulatory proteins interact with defined regions of the genome to render specific genes transcribable. While tenaciously bound as well as readily dissociable nonhistone chromosomal proteins have been purported to be the subclass of nonhistone chromosomal proteins which contains regulatory macromolecules, direct experimental evidence to distinguish between these two alternatives is at present limited. It is also presently unclear if nonhistone chromosomal proteins interact directly with DNA or with histone-DNA complexes.

In addition to understanding the mechanism by which nonhistone chromosomal proteins activate transcription of specific genes, the mechanism by which transcription is repressed must also be accounted for. One may envision inactivation of a gene or set of genes via degradation of the activator protein or proteins. Proteases which may utilize nonhistone chromosomal proteins as substrates have been shown to be associated with chromatin. An alternative mechanism for inactivation of regulatory proteins may involve acetate and/or phosphate groups added posttranslationally to nonhistone chromosomal proteins. One can speculate that repression of genetic sequences may be brought about by removal of such moieties from nonhistone chromosomal protein molecules. Deacetylases as well as phosphatases have been identified within the nucleus, lending credence to such speculation.

But perhaps most important, caution must be exercised in making unqualified generalizations regarding regulation of eukaryotic gene expression and eukaryotic gene regulators. Many of the genes thus far examined in an in-depth comprehensive manner, such as globin and ovalbumin genes, when expressed reflect a long-term commitment of a cell to expression of a differentiated gene product. These genes contain intervening sequences and their transcripts require a complex series of processing steps (cleavages, splicing, and chemical modifications) before they are translocated to the cytoplasm as viable templates for the synthesis of proteins. The type of transcriptional control operative in these situations and the nature of their regulators may be distinctively different from that associated with genes such as histone genes, or genes which code for metabolic enzymes, where expression is transient and acutely responsive to cellular requirements, e.g., DNA replication or substrate levels. From structural as well as functional standpoints it may be instructive to bear in mind that histone genes and genes which code for some metabolic enzymes do not contain intervening sequences and undergo a comparatively minimal amount of posttranscriptional processing. It would not be at all unrealistic to consider the possibility that such genes might be organized and regulated in a prokaryotic type manner.

As fractionation and characterization of the nonhistone chromosomal proteins progress, and additional information regarding the nature of sequences which are involved in control of specific genes becomes available the functional properties of eukaryotic gene regulators should become more apparent.

ACKNOWLEDGMENTS

We are indebted to Vicki Gates for her expert editorial assistance. Studies from this laboratory which are described in the chapter were supported by the following research grants: 5-217 from the National Birth Defects Foundation and PCM 80-18075 from the National Science Foundation.

REFERENCES

1. Olins, A. L. and Olins, D. E., Spheroid chromatin units (v bodies), *Science,* 183, 330, 1974.
2. Griffith, J. D., Chromatin structure: deduced from a minichromosome, *Science,* 187, 1202, 1975.
3. Oudet, P., Gross-Bellard, M., and Chambon, P., Electron microscope and biochemical evidence that chromatin structure is a repeating unit, *Cell,* 4, 281, 1967.
4. Pardon, J. F., Wilkins, M. H. F., and Richards, B. M., Super-helical model for nucleohistone, *Nature (London),* 215, 508, 1967.
5. Finch, J. T., Lutter, L. C., Rhodes, D., Brown, R. S., Rushton, B., Levitt, M., and Klug, A., Structure of nucleosome core particles, *Nature (London),* 269, 29, 1977.
6. Noll, M., Subunit structure of chromatin, *Nature (London),* 251, 249, 1974.
7. Shaw, B. R., Herman, T. M., Kovacic, R. T., Beaudreau, G. S., and Van Holde, K. E., Analysis of the subunit organization in chicken erythrocyte chromatin, *Proc. Natl. Acad. Sci. U.S.A.,* 73, 505, 1976.

8. **Todd, R. D. and Garrard, W. T.**, Overall pathway of mononucleosome production, *J. Biol. Chem.*, 254, 3074, 1979.

9. **Kornberg, R. D. and Thomas, J. O.**, Chromatin structure: oligomers of the histones, *Science*, 184, 865, 1974.

10. **Thomas, J. O. and Kornberg, R. D.**, An octamer of histones in chromatin and free in solution, *Proc. Natl. Acad. Sci. U.S.A.*, 72, 2626, 1975.

11. **Jackson, V.**, Studies on the histone organization in the nucleosome using formaldehyde as a reversible cross-linking agent, *Cell*, 15, 945, 1978.

12. **Mirzabekov, A. D., Shick, V. V., Belyavsky, A. V., and Bavykin, S. G.**, Primary organization of nucleosome core particle of chromatin sequence of histone arrangement along DNA, *Proc. Natl. Acad. Sci. U.S.A.*, 75, 4184, 1978.

13. **Mirzabekov, A. D., Shick, V. V., Belyavsky, A. V., Bavykin, S. G., and Rich, A.**, *Frontiers of Biorganic Chemistry and Molecular Biology (IUPAC)*, Symposium, Ananchenko, S. N., Ed., Pergamon Press, New York, 1980, 361.

14. **Noll, M., Thomas, J. O., and Kornberg, R. D.**, Preparation of native chromatin and damage caused by shearing, *Science*, 187, 1203, 1975.

15. **Lohr, D., Corden, J., Tachell, K., Kovacic, R. T., and Van Holde, K. E.**, Comparative subunit structure of HeLa, yeast and chicken erythrocyte chromatin, *Proc. Natl. Acad. Sci. U.S.A.*, 74, 79, 1977.

16. **Kornberg, R. D.**, Structure of chromatin, *Annu. Rev. Biochem.*, 46, 931, 1977.

17. **Whitlock, J. P., Jr. and Simpson, R. T.**, Removal of histone H1 exposes a fifty base pair DNA segment between nucleosomes, *Biochemistry*, 15, 3307, 1976.

18. **Noll, M. and Kornberg, R. D.**, Action of micrococcal nuclease on chromatin and the location of histone H1, *J. Mol. Biol.*, 109, 393, 1977.

19. **Simpson, R. T.**, Structure of the chromatosome, a chromatin particle containing 160 base pairs of DNA and all the histones, *Biochemistry*, 17, 5524, 1978.

20. **Weintraub, H.**, The nucleosomal repeat length increases during erythropoiesis in the chick, *Nucl. Acids Res.*, 5, 1179, 1978.

21. **Brown, I. R.**, Postnatal appearance of a short DNA repeat length in neurons of the cerebral cortex, *Biochem. Biophys. Res. Commun.*, 84, 285, 1978.

22. **Ermini, M. and Kuenzle, C. C.**, The chromatin repeat length of cortical neurons shortens during early postnatal development, *FEBS Lett.*, 90, 167, 1978.

23. **Stratling, W. H.**, Role of histone H1 in the conformation of oligonucleosomes as a function of ionic strength, *Biochemistry*, 18, 596, 1979.

24. **Worcel, A. and Benyajati, C.**, Higher order coiling of DNA in chromatin, *Cell*, 12, 83, 1977.

25. **Brasch, K., Seligy, V. L., and Setterfield, G.**, Effects of low salt concentration on structural organization and template activity of chromatin in chicken erythrocyte nuclei, *Exp. Cell Res.*, 65, 61, 1971.

26. **Finch, J. T. and Klug, A.**, Solenoidal model for superstructure in chromatin, *Proc. Natl. Acad. Sci. U.S.A.*, 73, 1897, 1976.

27. **Thoma, F. and Koller, Th.**, Influence of histone H1 on chromatin structure, *Cell*, 12, 101, 1977.

28. **Campbell, A. M., Cotter, R. I., and Pardon, J. F.**, Light scattering measurements supporting helical structures for chromatin in solution, *Nucl. Acids Res.*, 5, 1571, 1978.

29. **Lawn, R. M., Fritsch, E. F., Parker, R. C., Blake, G., and Maniatis, T.**, The isolation and characterization of linked δ- and β-globin genes from a cloned library of human DNA, *Cell*, 15, 1157, 1978.

30. **Thoma, F., Koller, Th., and Klug, A.**, Involvement of histone H1 in the organization of the nucleosome and the salt dependent superstructures of chromatin, *J. Cell Biol.*, 83, 403, 1979.

31. **McGhee, J. D., Rau, D. C., Charney, E., and Felsenfeld, G.**, Orientation of the nucleosome within the higher order structure of chromatin, *Cell*, 22, 87, 1978.

32. **Stratling, W. H., Muller, U., and Zentgraf, H.**, The higher order repeat structure of chromatin is built up of globular particles containing eight nucleosomes, *Exp. Cell. Res.*, 117, 301, 1979.

33. **Butt, T. R., Jump, D. B., and Smulson, M. E.**, Nucleosome periodicity in HeLa cell chromatin as probed by micrococcal DNase, *Proc. Natl. Acad. Sci. U.S.A.*, 76, 1628, 1979.

34. **Fujimoto, M., Kalinski, A., Pritchard, A. E., Kowalsi, D., and Laskowski, M.**, Accessibility of some regions of DNA in chromatin (chick erythrocyte) to single strand-specific nucleases, *J. Biol. Chem.*, 254, 7405, 1979.

35. **Renz, M., Nehls, P., and Hozier, J.**, Involvement of histone H1 in the organization of the chromosome fiber, *Proc. Natl. Acad. Sci. U.S.A.*, 74, 1879, 1977.

36. **MacGillivray, A. J., Cameron, A., Krauze, R. J., Rickwood, D., and Paul, J.**, The non-histone proteins of chromatin. Their isolation and composition in a number of tissues, *Biochim. Biophys. Acta*, 277, 384, 1972.

37. **Shaw, L. M. J. and Huang, R. C.**, A description of two procedures which avoid the use of extreme pH conditions for the resolution of components isolated from chromatins prepared from pig cerebellar and pituitary nuclei, *Biochemistry*, 9, 4530, 1970.

38. **Stein, G. and Baserga, R.,** The synthesis of acidic nuclear proteins in the prereplicative phase of the isoproterenol-stimulated salivary gland, *J. Biol. Chem.,* 245, 6097, 1970.

39. **Hill, R. J., Poccia, D. L., and Doty, P.,** Towards a total macromolecular analysis of sea urchin embryo chromatin, *J. Mol. Biol.,* 61, 445, 1971.

40. **Shelton, K. R. and Neelin, J. M.,** Nuclear residual proteins from goose erythroid cells and liver, *Biochemistry,* 10, 2342, 1971.

41. **Rovera, G. and Baserga, R.,** Early changes in the synthesis of acidic nuclear proteins in human diploid fibroblasts stimulated to synthesize DNA by changing the medium, *J. Cell. Physiol.,* 77, 201, 1971.

42. **Tsuboi, A. and Baserga, R.,** Synthesis of nuclear acidic proteins in density-inhibited fibroblasts stimulated to proliferate, *J. Cell. Physiol.,* 80, 107, 1972.

43. **Levy, R., Levy, S., Rosenberg, S. A., and Simpson, R. J.,** Selective stimulation of nonhistone chromatin protein synthesis in lymphoid cells by phytohemagglutinin, *Biochemistry,* 12, 224, 1973.

44. **LeStourgeon, W. M. and Rusch, H. P.,** Nuclear acid protein changes during differentiation in *Physarum polycephalum, Science,* 174, 1233, 1971.

45. **Karn, G., Johnson, E. M., Vidali, G., and Allfrey, V. G.,** Differential phosphorylation and turnover of nuclear acidic proteins during the cell cycle of synchronized HeLa cells, *J. Biol. Chem.,* 249, 667, 1974.

46. **Gronow, M. and Thackrah, T.,** The nonhistone nuclear proteins of some rat tissues, *Arch. Biochem. Biophys.,* 158, 337, 1973.

47. **Mullins, D. W., Jr., Giri, C. P., and Smulson, M.,** Poly(ADP) polymerase: the distribution of a chromosome-associated enzyme within the chromatin substructure, *Biochemistry,* 16, 506, 1977.

48. **Kleinsmith, L. J. and Allfrey, V. G.,** Nuclear phosphoproteins. I. Isolation and characterization of a phosphoprotein fraction from calf thymus nuclei, *Biochim. Biophys. Acta,* 175, 123, 1969.

49. **Benjamin, W. and Goodman, R. M.,** Phosphorylation of dipteran chromosomes and rat liver nuclei, *Science,* 166, 629, 1969.

50. **Ahmed, K.,** Studies of nuclear phosphoproteins of rat ventral prostate: incorporation of ^{32}P from [^{32}P]ATP, *Biochim. Biophys. Acta,* 243, 38, 1971.

51. **Rickwood, D., Riches, P. G., and MacGillivray, A. J.,** Studies of the *in vitro* phosphorylation of chromatin non-histone proteins in isolated nuclei, *Biochim. Biophys. Acta,* 299, 162, 1973.

52. **Phillips, I. R. and Mathias, A. P.,** Tissue specificity and phosphorylation of nonhistone chromosomal proteins studied by two-dimensional gel electrophoresis, in preparation.

53. **Thomson, J. A., Chiu, J. F., and Hnilica, L. S.,** Nuclear phosphoprotein kinase activities in normal and neoplastic tissues, *Biochim. Biophys. Acta,* 407, 114, 1975.

54. **Phillips, I. R., Shephard, E. A., Stein, J. L., Kleinsmith, L. J., and Stein, G. S.,** Nuclear protein kinase activities during the cell cycle of HeLa S$_3$ cells, *Biochim. Biophys. Acta,* 565, 326, 1979.

55. **Frenster, J. H., Allfrey, V. G., and Mirsky, A. E.,** Repressed and active chromatin isolated from interphase lymphocytes, *Proc. Natl. Acad. Sci. U.S.A.,* 50, 1026, 1963.

56. **Dingman, W. C. and Sporn, M. B.,** Studies on chromatin. I. Isolation and characterization of nuclear complexes of deoxyribonucleic acid, ribonucleic acid, and protein from embryonic and adult tissues of the chicken, *J. Biol. Chem.,* 239, 3483, 1964.

57. **Frenster, J. H.,** Nuclear polyanions as de-repressors of synthesis of ribonucleic acid, *Nature (London),* 206, 680, 1965.

58. **Spelsberg, T. C., Steggles, A. W., Chytil, F., and O'Malley, B. W.,** Progesterone-binding components of chick oviduct. V. Exchange of progesterone-binding capacity from target to nontarget tissue chromatins, *J. Biol. Chem.,* 247, 1368, 1972.

59. **Elgin, S. C. R. and Bonner, J.,** Limited heterogeneity of the major nonhistone chromosomal proteins, *Biochemistry,* 9, 4440, 1970.

60. **Seale, R. L. and Aronson, A. I.,** Chromatin-associated proteins of the developing sea urchin embryo. I. Kinetics of synthesis and characterization of non-histone proteins, *J. Mol. Biol.,* 75, 633, 1973.

61. **Yeoman, L. C., Taylor, C. W., Jordan, J. J., and Busch, H.,** Two-dimensional polyacrylamide gel electrophoresis of chromatin proteins of normal rat liver and Novikoff hepatoma ascites cells, *Biochem. Biophys. Res. Commun.,* 53, 1067, 1973.

62. **Teng, C. S., Teng, C. T., and Allfrey, V. G.,** Studies of nuclear acidic proteins. Evidence for their phosphorylation, tissue specificity, selective binding to deoxyribonucleic acid, and stimulatory effects on transcription, *J. Biol. Chem.,* 246, 3597, 1971.

63. **Spelsberg, T. C., Mitchell, W. M., Chytil, F., Wilson, E. M., and O'Malley, B. W.,** Chromatin of the developing chick oviduct: changes in the acidic proteins, *Biochim. Biophys. Acta,* 312, 765, 1973.

64. **Van den Broek, H. W. J., Nooden, L. D., Sevall, J. S., and Bonner, J.,** Isolation purification, and fractionation of nonhistone chromosomal proteins, *Biochemistry,* 12, 229, 1973.

65. **Richter, K. H. and Sekeris, C. E.,** Isolation and partial purification of non-histone chromosomal proteins from rat liver, thymus and kidney, *Arch. Biochem. Biophys.,* 148, 44, 1972.

66. **MacGillivray, A. J., Carroll, D., and Paul, J.,** The heterogeneity of the non-histone chromatin proteins from mouse tissues, *FEBS Letts.,* 13, 204, 1971.

67. **Gonzalez-Mujica, F. and Mathias, A. P.**, Proteins from different classes of liver nuclei in normal and thioacetamide-treated rats, *Biochem. J.*, 133, 441, 1973.

68. **Platz, R. D., Stein, G. S., and Kleinsmith, L. J.**, Changes in the phosphorylation of non-histone chromatin proteins during the cell cycle of HeLa S₃ cells, *Biochem. Biophys. Res. Commun.*, 51, 735, 1973.

69. **Platz, R. D. and Hnilica, L. S.**, Phosphorylation of nonhistone chromatin proteins during sea urchin development, *Biochem. Biophys. Res. Commun.*, 54, 222, 1973.

70. **Loeb, J. and Creuzet, C.**, Electrophoretic comparison of acidic proteins of chromatin from different animal tissues, *FEBS Lett.*, 5, 37, 1969.

71. **Wu, F. C., Elgin, S. C. R., and Hood, L. E.**, Nonhistone chromosomal proteins of rat tissues. A comparative study by gel electrophoresis, *Biochemistry*, 12, 2792, 1973.

72. **Elgin, S. C. R., Boyd, J. B., Hood, L. E., Wray, W., and Wu, F. C.**, A prologue to the study of the nonhistone chromosomal proteins, *Cold Spring Harbor Symp. Quant. Biol.*, 38, 821, 1974.

73. **Tsitilou, S., and Mathias, A. P.**, Proteins from different classes of nuclei in developing rat brain, in press.

74. **Catino, J. J., Yeoman, L. C., Mandel, M., and Busch, H.**, Characterization of DNA binding protein from rat liver chromatin which decreases during growth, *Biochemistry*, 17, 983, 1978.

75. **Barrett, T. and Gould, H. J.**, Tissue and species specificity of non-histone chromatin proteins, *Biochim. Biophys. Acta*, 294, 165, 1973.

76. **MacGillivray, A. J. and Rickwood, D.**, The heterogeneity of mouse-chromatin nonhistone proteins as evidenced by two-dimensional polyacrylamide gel electrophoresis and ion-exchange chromatography, *Eur. J. Biochem.*, 41, 181, 1974.

77. **Peterson, J. L. and McConkey, E. H.**, Non-histone chromosomal proteins from HeLa cells. A survey by high resolution, two-dimensional electrophoresis, *J. Biol. Chem.*, 251, 548, 1976.

78. **Peterson, J. L. and McConkey, E. H.**, Proteins of Friend leukemia cells. Comparison of hemoglobin-synthesizing and noninduced populations, *J. Biol. Chem.*, 251, 555, 1976.

79. **O'Farrell, P. H.**, High resolution two-dimensional electrophoresis of proteins, *J. Biol. Chem.*, 250, 4007, 1975.

80. **Goodwin, G. H., Shooter, K. V., and Johns, E. W.**, Interaction of a non-histone chromatin protein (high-mobility group protein 2) with DNA, *Eur. J. Biochem.*, 54, 427, 1975.

81. **Patel, G. L. and Thomas, T. L.**, Some binding parameters of chromatin acidic proteins with high affinity for deoxyribonucleic acid, *Proc. Natl. Acad. Sci. U.S.A.*, 70, 2524, 1973.

82. **Kleinsmith, L. J., Heidema, J., and Carroll, A.**, Specific binding of rat liver nuclear proteins to DNA, *Nature (London)*, 226, 1025, 1970.

83. **Kleinsmith, L. J.**, Specific binding of phosphorylated non-histone chromatin protein to deoxyribonucleic acid, *J. Biol. Chem.*, 248, 5648, 1973.

84. **Wakabayashi, K., Wang, S., Hord, G., and Hnilica, L. S.**, Tissue-specific nonhistone chromatin proteins with affinity for DNA, *FEBS Lett.*, 32, 46, 1973.

85. **Thomas, T. L. and Patel, G. L.**, DNA unwinding component of the nonhistone chromatin proteins, *Proc. Natl. Acad. Sci. U.S.A.*, 73, 4364, 1976.

86. **Allfrey, V. G.**, DNA-binding proteins and transcriptional control in prokaryotic and eukaryotic systems, in *Acidic Proteins of the Nucleus*, Cameron, I. L. and Jeter, J. R., Jr., Eds., Academic Press, New York, 1974.

87. **Allfrey, V. G., Inoue, A., Karn, J., Johnson, E. M., and Vidali, G.**, Phosphorylation of DNA-binding nuclear acidic proteins and gene activation in the HeLa cell cycle, *Cold Spring Harbor Symp. Quant. Biol.*, 38, 785, 1974.

88. **Allfrey, V. G., Inoue, A., and Johnson, E. M.**, Use of DNA columns to separate and characterize nuclear nonhistone proteins, in *Chromosomal Proteins and their Role in the Regulation of Gene Expression*, Stein, G. S. and Kleinsmith, L. J., Eds., Academic Press, New York, 1975, 265.

89. **Patel, G. L. and Thomas, T. L.**, Interactions of a subclass of nonhistone chromatin proteins with DNA, in *Chromosomal Proteins and their Role in the Regulation of Gene Expression*, Stein, G. S. and Kleinsmith, L. J., Eds., Academic Press, New York, 1975, 249.

90. **Sevall, J. S., Cockburn, A., Savage, M., and Bonner, J.**, DNA-protein interactions of the rat liver non-histone chromosomal protein, *Biochemistry*, 14, 782, 1975.

91. **Champoux, J. J. and McConaughy, B. L.**, Purification and characterization of the DNA untwisting enzyme from rat liver, *Biochemistry*, 14, 782, 1975.

92. **Goodwin, G. H., Nicolas, R. H., and Johns, E. W.**, Microheterogeneity in a non-histone chromosomal protein, *FEBS Lett.*, 64, 412, 1976.

93. **Kruh, J., Tichonicky, L., and Wajcman, H.**, Action des proteines acides du noyau sur la synthesis acelulaire de l'hemoglobine et sur les RNA, *Biochim. Biophys. Acta*, 196, 549, 1969.

94. **Platz, R. D., Kish, V. M., and Kleinsmith, L. J.**, Tissue specificity of nonhistone chromatin phosphoproteins, *FEBS Lett.*, 12, 38, 1970.

95. **Wang, T. Y.**, Tissue specificity of non-histone chromosomal proteins, *Exp. Cell Res.*, 69, 217, 1971.

96. **Wilhelm, J. A., Ansevin, A. T., Johnson, A. W., and Hnilica, L. S.**, Proteins of chromatin in genetic restriction. IV. Comparison of histone and nonhistone proteins of rat liver nucleolar and extranu-cleolar chromatin, *Biochim. Biophys. Acta,* 272, 220, 1972.

97. **Bekhor, I., Anne, L., Kim, J., Lapeyre, J.-N., and Stambaugh, R.**, Organ discrimination through organ-specific nonhistone chromosomal proteins, *Arch. Biochem. Biophys.,* 161, 11, 1974.

98. **Davis, R. H., Wilson, R. B., and Ebadi, M. S.**, Tissue specificity of nuclear acidic proteins isolated from bovine brain and adrenal medulla, *Can. J. Biochem.,* 53, 101, 1975.

99. **Fujitani, H. and Holoubek, V.**, Presence of the same types of nonhistone chromosomal proteins of different tissues, *Experimentia,* 30, 474, 1974.

100. **Chytil, F. and Spelsberg, T. C.**, Tissue differences in antigenic properties of non-histone protein-DNA complexes, *Nature New Biol.,* 233, 215, 1971.

101. **Zardi, L., Lin, J. C., and Baserga, R.**, Immunospecificity to non-histone chromosomal proteins of anti-chromatin antibodies, *Nature New Biol.,* 245, 211, 1973.

102. **Wakabayashi, K. and Hnilica, L. S.**, The immunospecificity of nonhistone protein complexes with DNA, *Nature New Biol.,* 242, 153, 1973.

103. **Chiu, J. F., Craddock, C., Morris, H. P., and Hnilica, L. S.**, Immunospecificity of chromatin non-histone protein-DNA complexes in normal and neoplastic growth, *FEBS Lett.,* 42, 94, 1974.

104. **Wakabayashi, K., Wang, S., and Hnilica, L. S.**, Immunospecificity of nonhistone proteins in chro-matin, *Biochemistry,* 13, 1027, 1974.

105. **Wang, S., Chiu, J. F., Klyszejko-Stefanowicz, L., Fujitani, H., Hnilica, L. S., and Ansevin, A. T.**, Tissue-specific chromosomal non-histone protein interactions with DNA, *J. Biol. Chem.,* 5, 1471, 1976.

106. **Wu, F. C., Elgin, S. C. R., and Hood, L. E.**, The nonhistone chromosomal proteins of vertebrate liver and kidney: a comparative study by gel electrophoresis, *J. Mol. Evol.,* 5, 87, 1975.

107. **Shooter, K. V., Goodwin, G. H., and Johns, E. W.**, Interactions of a purified non-histone chromo-somal protein with DNA and histone, *Eur. J. Biochem.,* 47, 263, 1974.

108. **Yu, S. S., Li, H. J., Goodwin, G. H., and Johns, E. W.**, Interaction of non-histone chromosomal proteins HMG1 and HMG2 with DNA, *Eur. J. Biochem.,* 78, 497, 1977.

109. **Fujitani, H. and Holoubek, V.**, Nonhistone nuclear proteins of rat brain, *J. Neurochem.,* 23, 1215, 1974.

110. **Olpe, H. R., von Hahn, H. P., and Honegger, C. G.**, Differences in electrophoretic patterns of non-histone proteins from large and small nuclei of rat brain, *Brain Res.,* 58, 453, 1973.

111. **Tashiro, T., Mizobe, F., and Kurokawa, M.**, Characteristics of cerebral non-histone chromatin pro-teins as revealed by polyacrylamide gel electrophoresis, *FEBS Lett.,* 38, 121, 1974.

112. **Fleischer-Lambropoulos, H., Sarkander, H. I., and Brade, W. P.**, Phosphorylation of nonhistone chromatin proteins from neuronal and glial nuclei-enriched fractions of rat brain, *FEBS Lett.,* 45, 329, 1974.

113. **Mirsky, A. E. and Ris, H.**, The composition and structure of isolated chromosomes, *J. Gen. Physiol.,* 34, 475, 1951.

114. **Busch, H., Ballal, N. R., Olson, M. O. J., and Yeoman, L. C.**, Chromatin and its nonhistone pro-teins, in *Methods in Cancer Research,* Vol. 11, Busch, H., Ed., Academic Press, New York, 1975, 43.

115. **Ozaki, H.**, Developmental studies of sea urchin chromatin. Chromatin isolated from spermatozoa of the sea urchin *Strongylocentrotus purpuratus, Dev. Biol.,* 26, 209, 1971.

116. **Mohberg, J. and Rusch, H. P.**, Nuclear histones in *Physarum polycephalum, Arch. Biochem. Bio-phys.,* 138, 418, 1970.

117. **Gershey, E. L. and Kleinsmith, L. J.**, Phosphorylation of nuclear proteins in avian erythrocytes, *Biochim. Biophys. Acta,* 194, 519, 1969.

118. **Marushige, K. and Dixon, G. H.**, Developmental changes in chromosomal composition and template activity during spermatogenesis in trout testis, *Dev. Biol.,* 19, 397, 1969.

119. **Marushige, K. and Ozaki, H.**, Properties of isolated chromatin from sea urchin embryo, *Dev. Biol.,* 16, 474, 1967.

120. **Dolbeare, F. and Koenig, H.**, Fractionation of rat liver chromatin: effects of cations, hepatectomy and actinomycin D, *Proc. Soc. Exp. Biol. Med.,* 135, 636, 1970.

121. **Marushige, K. and Bonner, J.**, Fractionation of liver chromatin, *Proc. Natl. Acad. Sci. U.S.A.,* 68, 2941, 1971.

122. **Reeck, G. R., Simpson, R. T., and Sober, H. A.**, Resolution of a spectrum of nucleoprotein species in sonicated chromatin, *Proc. Natl. Acad. Sci. U.S.A.,* 69, 2317, 1972.

123. **Berkowitz, E. M. and Doty, P.**, Chemical and physical properties of fractionated chromatin, *Proc. Natl. Acad. Sci. U.S.A.,* 72, 3328, 1975.

124. **Neelin, J. M., Mazen, A., and Champagne, M.**, The fractionation of active and inactive chromatins from erythroid cells of chicken, *FEBS Lett.,* 65, 309, 1976.

125. **Rodriguez, L. V. and Becker, F. F.**, Rat liver chromatin. Distribution of histones and nonhistone proteins in eu- and heterochromatin, *Arch. Biochem. Biophys.,* 173, 438, 1976.

126. Comings, D. E., Harris, D. C. Okada, T. A., and Holmquist, G., Nuclear proteins. IV. Deficiency of non-histone proteins in condensed chromatin of *Drosophila virilis* and mouse, *Exp. Cell Res.,* 105, 349, 1977.

127. Simpson, R. T. and Reeck, G. R., A comparison of the proteins of condensed and extended chromatin factors of rabbit liver and calf thymus, *Biochemistry,* 12, 3853, 1973.

128. Murphy, E. C., Hall, S. H., Shepherd, J. H., and Weiser, R. S., Fractionation of mouse myeloma chromatin, *Biochemistry,* 12, 3843, 1973.

129. Gottesfeld, J. M., Garrard, W. T., Bagi, G., Wilson, R. F., and Bonner, J., Partial purification of the template-active fraction of chromatin: a preliminary report, *Proc. Natl. Acad. Sci. U.S.A.,* 71, 2193, 1974.

130. Comings, D. E. and Harris, D. C., Nuclear proteins. I. Electrophoretic comparison of mouse nucleoli, heterochromatin, euchromatin and contractile proteins, *Exp. Cell Res.,* 96, 161, 1975.

131. Musich, P. R., Brown, F. L., and Maio, J. J., Subunit structure of chromatin and the organization of eukaryotic highly repetitive mammalian DNA, *Proc. Natl. Acad. Sci. U.S.A.,* 74, 3297, 1975.

132. Chiu, N., Baserga, R., and Furth, J. J., Composition and template activity of chromatin fractionated by isoelectric focusing, *Biochemistry,* 16, 4796, 1977.

133. Zardi, L., Chicken antichromatin antibodies: specificity to different chromatin fractions, *Eur. J. Biochem.,* 55, 231, 1975.

134. Keller, R. K., Socher, S. H., Krall, J. F., Chandra, T., and O'Malley, B. W., Fractionation of chick oviduct chromatin. IV. Association of protein kinase with transcriptionally active chromatin, *Biochem. Biophys. Res. Commun.,* 66, 453, 1975.

135. Gottesfeld, J. M. and Partington, G. A., Distribution of messenger RNA-coding sequences in fractionated chromatin, *Cell,* 12, 953, 1977.

136. Wallace, R. B., Dube, S. K., and Bonner, J., Localization of the globin gene in the template active fraction of chromatin of Friend leukemia cell, *Science,* 198, 1166, 1977.

137. Krieg, P. and Wells, J. R. E., The distribution of active genes (globin) and inactive genes (keratin) in fractionated chicken erythroid chromatin, *Biochemistry,* 15, 4549, 1976.

138. Rovera, G. and Baserga, R., Effect of nutritional changes in chromatin template activity and non-histone chromosomal protein synthesis in WI-38 and 3T6 cells, *Exp. Cell Res.,* 78, 118, 1973.

139. Johnson, E. M., Karn, J., and Allfrey, V. G., Early nuclear events in the induction of lymphocyte proliferation by mitogens. Effects of concanavalin A on the phosphorylation and distribution of non-histone chromatin proteins, *J. Biol. Chem.,* 249, 4990, 1974.

140. Decker, J. M. and Marchalonis, J. J., Molecular events in lymphocyte differentiation: stimulation of nonhistone nuclear protein synthesis in rabbit peripheral blood lymphocytes by anti-immunoglobulin, *Biochem. Biophys. Res. Commun.,* 74, 584, 1977.

141. Hemminki, K., Synthesis of chromatin proteins in resting and stimulated human lymphocyte populations, *Exp. Cell Res.,* 93, 63, 1975.

142. Garrard, W. T. and Bonner, J., Changes in chromatin proteins during liver regeneration, *J. Biol Chem.,* 249, 5570, 1974.

143. Cameron, I. L., Griffin, E. E., and Rudick, M. J., Macromolecular events following refeeding of starved *Tetrahymena, Exp. Cell Res.,* 65, 262, 1971.

144. Rudick, M. J. and Cameron, I. L., Regulation of DNA synthesis and cell division in starved-refed synchronized *Tetrahymena pyriformis* HSM, *Exp. Cell Res.,* 70, 411, 1972.

145. Jeter, J. R., Jr., Pavlat, W. A., and Cameron, I. L., Changes in the nuclear acidic proteins and chromatin structure in starved and refed *Tetrahymena, Exp. Cell Res.,* 93, 79, 1975.

146. Weisenthal, L. M. and Ruddon, R. W., Characterization of human leukemia and Burkitt lymphoma cells by their acidic nuclear protein profiles, *Cancer Res.,* 32, 1009, 1972.

147. Zornetzer, M. S. and Stein, G. S., Gene expression in mouse neuroblastoma cells: properties of the genome, *Proc. Natl. Acad. Sci. U.S.A.,* 72, 3119, 1975.

148. Yeoman, L. C., Taylor, C. W., Jordan, J. J., and Busch, H., Differences in chromatin proteins of growing and non-growing tissues, *Exp. Cell Res.,* 91, 207, 1975.

149. Yeoman, L. C., Seeber, S., Taylor, C. W., Fernbach, D. J., Falletta, J. M., Jordan, J. J., and Busch, H., Differences in chromatin proteins of resting and growing human lymphocytes, *Exp. Cell Res.,* 100, 47, 1976.

150. Sevaljevic, L., Popovic, Z., and Konstantinovic, M., Investigation on stage-related changes of sea urchin chromatin proteins by hydroxyapatite chromatography, *Int. J. Biochem.,* 6, 903, 1975.

151. Sevaljevic, L., Krtolica, K., and Konstantinovic, M., Embryonic stage-related properties of sea urchin embryo chromatin, *Biochim. Biophys. Acta,* 425, 76, 1976.

152. Theriault, J. and Landesman, R., An analysis of acidic nuclear proteins during the development of *Xenopus laevis, Cell Differ.,* 3, 249, 1974.

153. Teng, C. S., Nuclear acidic protein of the developing *Oncopeltus* embryos, *Biochim. Biophys. Acta,* 366, 385, 1974.

154. Vidali, G., Boffa, L. C., Littau, V. C., Allfrey, K. M., and Allfrey, V. G. Changes in nuclear acidic protein complement of red blood cells during embryonic development, *J. Biol. Chem.*, 248, 4065, 1973.

155. Pipkin, J. L., Jr. and Larson, D. A., Changing patterns of nucleic acids, basic and acidic proteins in generative and vegetative nuclei during pollen germination and pollen tube growth in *Hippeastrum belladonna*, *Exp. Cell Res.*, 79, 28, 1973.

156. Spivak, J. L., Chromosomal protein synthesis during erythropoiesis in the mouse spleen, *Exp. Cell Res.*, 91, 253, 1975.

157. Keppel, F., Allet, B., and Eisen, H., Appearance of a chromatin protein during the erythroid differentiation of Friend virus-transformed cells, *Proc. Natl. Acad. Sci. U.S.A.*, 74, 653, 1977.

158. Kadohama, N. and Turkington, R. W., Changes in acidic chromatin proteins during the hormone-dependent development of rat testis and epididymis, *J. Biol. Chem.*, 249, 6225, 1974.

159. Platz, R. D., Grimes, S. R., Hord, G., Meistrich, M. L., and Hnilica, L. S., Changes in nuclear proteins during embryonic development and cellular differentiation, in *Chromosomal Proteins and their Role in the Regulation of Gene Expression*, Stein, G. S. and Kleinsmith, L. J., Eds., Academic Press, New York, 1975, 67.

160. Allfrey, V. G., Johnson, E. M., Karn, J., and Vidali, G., Phosphorylation of nuclear proteins at times of gene activation, in *Protein Phosphorylation in Control Mechanisms*, Miami Winter Symposia, Vol. 5, Huijing, F. and Lee, E. Y. C., Eds., Academic Press, New York, 1973, 217.

161. Lazo, J. S., Prasad, K. N., and Ruddon, R. W., Synthesis and phosphorylation of chromatin-associated proteins in cAMP-induced "differentiated" neuroblastoma cells in culture, *Exp. Cell Res.*, 100, 41, 1976.

162. Banks-Schlegel, S., Martin, T. E., and Moscona, A. A., Synthesis and phosphorylation of chromosomal nonhistone proteins during embryonic development of neural retina, *Dev. Biol.*, 50, 1, 1976.

163. Burdman, J. A., The relationship between DNA synthesis and the synthesis of nuclear proteins in rat brain during development, *J. Neurochem.*, 19, 1459, 1972.

164. Buck, M. D. and Schauder, P., *In vivo* stimulation of 14C-amino acid incorporation into nonhistone proteins in rat liver chromatin induced by insulin and cortisol, *Biochim. Biophys. Acta*, 224, 644, 1970.

165. Swaneck, G. E., Chu, L., and Edelman, I., Stereospecific binding of aldosterone to renal chromatin, *J. Biol. Chem.*, 245, 5382, 1970.

166. Ruddon, R. W. and Rainey, C. H., Stimulation of nuclear protein synthesis in rat liver after phenobarbital administration, *Biochem. Biophys. Res. Commun.*, 40, 152, 1970.

167. Teng, C. S. and Hamilton, T. H., Regulation by estrogen of organ-specific synthesis of a nuclear acidic protein, *Biochem. Biophys. Res. Commun.*, 40, 1231, 1970.

168. Cohen, M. E. and Hamilton, T. H., Effect of estradiol-17b on the synthesis of specific uterine non-histone chromosomal proteins, *Proc. Natl. Acad. Sci. U.S.A.*, 72, 4346, 1975.

169. Hemminki, K., Labelling of oviduct nuclear and nucleolar proteins during estrogen induced differentiation, *Mol. Cell. Biochem.*, 11, 9, 1976.

170. Mainwaring, W. I. P., Rennie, P. S., and Keen, J., The androgenic regulation of prostate proteins with a high affinity for deoxyribonucleic acid. Evidence for a prostate deoxyribonucleic acid-unwinding protein, *Biochem. J.*, 156, 253, 1976.

171. Alonso, A. and Arnold, H. P., Stimulation of amino acid incorporation into rat liver nonhistone chromatin proteins after treatment with diethylnitrosamine, *FEBS Lett.*, 41, 8, 1974.

172. Kadohama, N. and Turkington, R. W., Altered populations of acidic chromatin proteins in breast cancer cells, *Cancer Res.*, 33, 1194, 1973.

173. Orrick, L. R., Olson, M. O. J., and Busch, H., Comparison of nucleolar proteins of normal rat liver and Novikoff hepatoma ascites cells by two-dimensional polyacrylamide gel electrophoresis, *Proc. Natl. Acad. Sci. U.S.A.*, 70, 1316, 1973.

174. Chae, C. B., Smith, M. C., and Morris, H. P., Chromosomal nonhistone proteins of rat hepatomas and normal rat liver, *Biochem. Biophys. Res. Commun.*, 60, 1468, 1974.

175. Boffa, L. C., Vidali, G., and Allfrey, V. G., Selective synthesis and accumulation of nuclear non-histone proteins during carcinogenesis of the colon induced by 1,2-dimethylhydrazine, *Cancer*, 36, 2356, 1975.

176. Boffa, L. C., Vidali, G., and Allfrey, V. G., Changes in nuclear non-histone protein composition during normal differentiation and carcinogenesis of intestinal epithelial cells, *Exp. Cell Res.*, 98, 396, 1976.

177. Biessman, H. and Rajewsky, M. F., Nuclear protein patterns in developing and adult brain and in ethylnitrosourea-induced neuroectodermal tumours of the rat, *J. Neurochem.*, 24, 387, 1975.

178. Biessman, H. and Rajewsky, M. F., The synthesis of brain chromosomal proteins after a pulse of the nervous system-specific carcinogen N-ethyl-N-nitrosourea to the fetal rat, *J. Neurochem.*, 27, 927, 1976.

179. Kadohama, N. and Anderson, K. M., Nuclear non-histone proteins from rat ventral prostate cells undergoing hypertrophy or hyperplasia, *Exp. Cell Res.*, 99, 135, 1976.

180. Stein, G. S., Park, W. D., Stein, J. L., and Lieberman, M. W., Synthesis of nuclear proteins during DNA repair synthesis in human diploid fibroblasts damaged with ultraviolet radiation or N-acetoxy-2-acetylaminofluorene, *Proc. Natl. Acad. Sci. U.S.A.*, 73, 1466, 1976.

181. Rovera, G., Baserga, R., and Defendi, V., Early increase in nuclear acidic protein synthesis after SV-40 infection, *Nature New Biol.*, 237, 240, 1972.

182. Ledinko, L., Nuclear acidic protein changes in adenovirus-infected human embryo kidney cultures, *Virology*, 54, 294, 1973.

183. Krause, M. O. and Stein, G. S., Modifications in the chromosomal proteins of SV-40 transformed WI-38 human diploid fibroblasts, *Biochem. Biophys. Res. Commun.*, 59, 796, 1974.

184. Krause, M. O., Kleinsmith, L. J., and Stein, G. S., Properties of the genome in normal and SV40 transformed WI38 human diploid fibroblasts. I. Composition and metabolism of nonhistone chromosomal proteins, *Exp. Cell Res.*, 92, 164, 1975.

185. Krause, M. O., Kleinsmith, L. J., and Stein, G. S., Properties of the genome in normal and SV-40 transformed WI-38 human diploid fibroblasts. III. Turnover of nonhistone chromosomal proteins and their phosphate groups, *Life Sci.*, 16, 1047, 1975.

186. Iida, H. and Oda, K., Stimulation of nonhistone chromosomal protein synthesis in simian virus 40-infected simian cells, *J. Virol.*, 15, 471, 1975.

187. Stein, G. S., Moscovici, G., Moscovici, C., and Mons, M., Acidic nuclear protein synthesis in Rous sarcoma virus infected chick embryo fibroblasts, *FEBS Lett.*, 38, 295, 1974.

188. Gonzalez, C. A. and Rees, K. R., Non-histone chromosomal proteins from virus-transformed and untransformed 3T3 mouse fibroblasts, *Biochim. Biophys. Acta*, 395, 361, 1975.

189. Krause, M. O., Noonan, K. D., Kleinsmith, L. J., and Stein, G. S., The effect of SV40 transformation on the chromosomal proteins of 3T3 mouse embryo fibroblasts, *Cell Differ.*, 5, 83, 1976.

190. Johnson, T. C. and Holland, J. J., Ribonucleic acid and protein synthesis in mitotic HeLa cells, *J. Cell Biol.*, 27, 565, 1965.

191. Pfeiffer, S. E. and Tolmach, L. J., RNA synthesis in synchronously growing populations of HeLa S3 cells. I. Rate of total RNA synthesis and its relationship to DNA synthesis, *J. Cell Physiol.*, 71, 77, 1968.

192. Farber, J., Stein, G., and Baserga, R., The regulation of RNA synthesis during mitosis, *Biochem. Biophys. Res. Commun.*, 47, 790, 1972.

193. Zeilig, C. E., Johnson, R. A., Friedman, D. L., and Sutherland, E. W., Cyclic AMP concentrations in synchronized HeLa cells, *J. Cell Biol.*, 55, 296a, 1972.

194. Kleinsmith, L. J., Allfrey, V. G., and Mirsky, A. E., Phosphorylation of nuclear protein early in the course of gene activation in lymphocytes, *Science*, 154, 780, 1966.

195. Pogo, B. G. T. and Katz, J. R., Early events in lymphocyte transformation by phytohaemagglutinin. Synthesis and phosphorylation of nuclear proteins, *Differentiation*, 2, 119, 1974.

196. Horenstein, A., Piras, M. M., Mordoh, J., and Piras, R., Protein phosphokinase activities of resting and proliferating human lymphocytes, *Exp. Cell Res.*, 101, 260, 1976.

197. Ishida, H. and Ahmed, K., Studies on phosphoproteins of submandibular gland nuclei isolated from isoproterenol treated rats, *Exp. Cell Res.*, 78, 31, 1973.

198. Ishida, H. and Ahmed, K., Studies on chromatin protein phosphokinase of submandibular gland from isoproterenol treated rats, *Exp. Cell Res.*, 84, 127, 1974.

199. Rieber, M. and Bacalao, J., Alterations in nuclear phosphoproteins of a temperature-sensitive Chinese hamster cell line exposed to non-permissive conditions, *Exp. Cell Res.*, 85, 334, 1974.

200. Marty de Morales, M., Blat, C., and Harel, L., Changes in the phosphoralation of non-histone chromosomal proteins in relationship to DNA and RNA synthesis in $BHK_{21}C_{13}$ cells, *Exp. Cell Res.*, 86, 111, 1974.

201. Pumo, D. E., Stein, G. S., and Kleinsmith, L. J., Phosphorylation of nonhistone chromosomal proteins early during the prereplicative phase of the cell cycle of WI-38 human diploid fibroblasts, *Cell Differ.*, 5, 45, 1976.

202. Brade, W. P., Thomson, J. A., Chiu, J. F., and Hnilica, L. S., Chromatin-bound kinase activity and phosphorylation of chromatin non-histone proteins during early kidney regeneration after folic acid, *Exp. Cell Res.*, 84, 183, 1974.

203. Ezrailson, E. G., Olson, M. O. J., Guetzow, K. A., and Busch, H., Phosphorylation of non-histone chromatin proteins in normal and regenerating rat liver, Novikoff hepatoma and rat heart, *FEBS Lett.*, 62, 69, 1976.

204. Man, N. T., Morris, G. E., and Cole, R. J., Phosphorylation of nuclear proteins during muscle differentiation *in vitro*, *FEBS Lett.*, 42, 257, 1974.

205. Turkington, R. W. and Riddle, M., Hormone-dependent phosphorylation of nuclear proteins during mammary gland differentiation *in vitro*, *J. Cell Biol.*, 244, 6040, 1969.

206. Plamer, W. K., Castagna, M., and Walsh, D. A., Nuclear protein kinase activity in glucagon-stimulated perfused rat livers, *Biochem. J.*, 143, 469, 1974.

207. Bottoms, G. D. and Jungmann, R. A., Effect of corticosterone on phosphorylation of rat liver nuclear proteins *in vitro*, *Proc. Soc. Exp. Biol. Med.*, 144, 83, 1973.

208. Trajkovic, D., Ribarac-Stepic, N., and Kanazir, D., The effect of cortisol on the phosphorylation of rat liver nuclear acidic proteins and the role of these proteins in biosynthesis of nuclear RNA, *Arch. Int. Physiol. Biochim.,* 82, 211, 1974.

209. Schauder, P., Starman, B. J., and Williams, R. H., Effect of cortisol on phosphorylation of acidic proteins in liver nuclei from adrenalectomized rats, *Experimentia,* 30, 1277, 1974.

210. Ahmed, K. and Ishida, H., Effect of testosterone on nuclear phosphoproteins of rat ventral prostate, *Mol. Pharmacol.,* 7, 323, 1971.

211. Ahmed, K. and Wilson, J. J., Chromatin-associated protein phosphokinases of rat ventral prostate, *J. Biol. Chem.,* 250, 2370, 1975.

212. Schauder, P., Starman, B. J., and Williams, R. H., Effect of testosterone of phenol-soluble nuclear acidic proteins of rat ventral prostate, *Proc. Soc. Exp. Biol. Med.,* 145, 331, 1974.

213. Liew, C. C., Suria, D., and Gornall, A. G., Effects of aldosterone on acetylation and phosphorylation of chromosomal proteins, *Endocrinology,* 93, 1025, 1973.

214. Cohen, M. E. and Kleinsmith, L. J., Effect of estradiol-17β on the phosphorylation of uterine nonhistone chromosomal proteins, *Fed. Proc.,* 34, 704, 1975.

215. Jungmann, R. A. and Schweppe, J. S., Mechanism of action of gonadotropin. I. Evidence for gonadotropin-induced modifications of ovarian nuclear basic and acidic protein biosynthesis, phosphorylation, and acetylation, *J. Biol. Chem.,* 247, 5535, 1972.

216. Kruh, J. and Tichonicky, L., Effect of triiodothyronine on rat liver chromatin protein kinase, *Eur. J. Biochem.,* 62, 109, 1976.

217. Salem, R. and DeVellis, J., Protein kinase activity and cAMP-dependent protein phosphorylation in subcellular fractions after norepinephrine treatment of glial cells, *Fed. Proc.,* 35, 296, 1976.

218. Blankenship, J. and Bresnick, E., Effects of phenobarbital on the phosphorylation of acidic nuclear proteins of rat liver, *Biochim. Biophys. Acta,* 340, 218, 1974.

219. Chiu, J. F., Craddock, C., Getz, S., and Hnilica, L. S., Nonhistone chromatin protein phosphorylation during azo-dye carcinogenesis, *FEBS Lett.,* 33, 247, 1973.

220. Chiu, J. F., Brade, W. P., Thomson, J., Tsai, Y. H., and Hnilica, L. S., Non-histone protein phosphorylation in normal and neoplastic rat liver chromatin, *Exp. Cell Res.,* 91, 200, 1975.

221. Jost, E., Lennox, R., and Harris, H., Affinity chromatography of DNA-binding proteins from human, murine and man-mouse hybrid cell lines, *J. Cell Sci.,* 18, 41, 1975.

222. Ahmed, K., Increased phosphorylation of nuclear phosphoproteins in precancerous liver, *Res. Commun. Chem. Pathol. Pharmacol.,* 9, 771, 1974.

223. Brade, W. P., Chiu, J. F., and Hnilica, L. S., Phosphorylation of rat liver nuclear acidic phosphoproteins after administration of α-1,2,3,4,5,6-hexachlorocyclohexane *in vivo, Mol. Pharmacol.,* 10, 398, 1974.

224. Pumo, D. E., Stein, G. S., and Kleinsmith, L. J., Stimulated phosphorylation of non-histone phosphoproteins in SV-40 transformed WI-38 human diploid fibroblasts, *Biochim. Biophys. Acta,* 402, 125, 1975.

225. Farron-Furstenthal, F., Protein kinases in hepatoma, and adult and fetal liver of the rat. I. Subcellular distribution, *Biochem. Biophys. Res. Commun.,* 67, 307, 1975.

226. Stein, G. S., Stein, J. L., and Thomson, J. A., Chromosomal proteins in transformed and neoplastic cells: a review, *Cancer Res.,* 38, 1181, 1978.

227. Stein, G. S., Spelsberg, T. C., and Kleinsmith, L. J., Nonhistone chromosomal proteins and gene regulation, *Science,* 183, 817, 1974.

228. Spelsberg, T. C., The role of nuclear acidic proteins in binding steroid hormones, in *Acidic Proteins of the Nucleus,* Cameron, I. L. and Jeter, J. R., Jr., Eds., Academic Press, New York, 1974, 247.

229. Thrall, C. L., Webster, R. A., and Spelsberg, T. C., Steroid receptor interaction with chromatin, in *The Cell Nucleus,* Vol. 6, Busch, H., Ed., Academic Press, New York, 1979.

230. Spelsberg, T. C., Steggles, A. W., and O'Malley, B. W., Progesterone-binding components of chick oviduct. III. Chromatin acceptor sites, *J. Biol. Chem.,* 246, 4188, 1971.

231. Defer, M., Dastugue, B., and Kruh, J., Direct binding of corticosterone and estradiol to rat liver nuclear non-histone proteins, *Biochimie,* 56, 559, 1974.

232. Charles, M. A., Ryffel, G. U., Obinata, M., McCarthy, B. J., and Baxter, J. D., Nuclear receptors for thyroid hormone: evidence for nonrandom distribution within chromatin, *Proc. Natl. Acad. Sci. U.S.A.,* 72, 1787, 1975.

233. Klyzsejko-Stefanowicz, L., Chiu, J. F., Tsai, Y. H., and Hnilica, L. S., Acceptor proteins in rat and androgenic tissue chromatin, *Proc. Natl. Acad. Sci. U.S.A.,* 73, 1954, 1976.

234. Spelsberg, T. C., Webster, R., and Pikler, G. M., Multiple binding sites for progesterone in the hen oviduct nucleus: evidence that acidic proteins represent the acceptors, in *Chromosomal Proteins and their Role in the Regulation of Gene Expression,* Stein, G. S. and Kleinsmith, L. J., Eds., Academic Press, New York, 1975, 153.

235. Webster, R. A., Pikler, G. M., and Spelsberg, T. C., Nuclear binding of progesterone hen oviduct. Role of acidic chromatin proteins in high-affinity binding, *Biochem. J.,* 156, 409, 1976.

236. Spelsberg, T. C., Webster, R. A., and Pikler, G. M., Chromosomal proteins regulate steroid binding to chromatin, *Nature (London)*, 262, 65, 1976.

237. Spelsberg, T. C., Webster, R. A., Pikler, G. M., Thrall, C., and Wells, D., Role of nuclear proteins as high affinity sites ("acceptors") for progesterone in the avian oviduct, *J. Steroid Biochem.*, 7, 1091, 1976.

238. Spelsberg, T. C., Webster, R. A., Pikler, G. M., Thrall, C. L., and Wells, D. J., Nuclear binding sites ("acceptors") for progesterone in avian oviduct: characterization of the highest-affinity sites, *Ann. N.Y. Acad. Sci.*, 286, 43, 1977.

239. Spelsberg, T. C., Thrall, C. L., Webster, R. A., and Pikler, G. M., Hormone regulation of growth: stimulatory and inhibitory influences of estrogens on DNA synthesis, *J. Toxicol. Environ. Health*, 3, 309, 1977.

240. Schwartz, R. J., Kuhn, R. W., Buller, R. E., Schrader, W. T., and O'Malley, B. W., Progesterone-binding components of chick oviduct. *In vitro* effects of purified hormone receptor complexes on the initiation of RNA synthesis in chromatin, *J. Biol. Chem.*, 251, 5166, 1976.

241. Tsai, S. Y., Tsai, M. J., Harris, S. E., and O'Malley, B. W., Effects of estrogen on gene expression in the chick oviduct. Control of ovalbumin gene expression by non-histone proteins, *J. Biol. Chem.*, 251, 6475, 1976.

242. Herrick, G. and Alberts, B., Nucleic acid helix-coil transitions mediated by helix-unwinding proteins from calf thymus, *J. Biol. Chem.*, 251, 2133, 1976.

243. Herrick, G., Delius, H., and Alberts, B., Single-stranded DNA structure and DNA polymerase activity in the presence of nucleic acid helix-unwinding proteins from calf thymus, *J. Biol. Chem.*, 251, 2142, 1976.

244. Schauder, P., Starman, B. J., and Williams, R. H., Effect of testosterone of phenol-soluble nuclear acidic proteins of rat ventral prostate, *Proc. Soc. Exp. Biol. Med.*, 145, 331, 1974.

245. Clark, B. F. C., Towards a total human map, *Nature (London)*, 292, 491, 1981.

246. Garrels, J. J., Two dimensional gel electrophoresis and computer analysis of proteins synthesized by clonal cell lines, *J. Biol. Chem.*, 245, 7961, 1979.

247. O'Farrel, P. Z., Goodman, H. M., and O'Farrell, P. H., High resolution two dimensional electrophoresis of basic as well as acidic proteins, *Cell*, 12, 1133, 1977.

248. Stein, J. L., Stein, G. S., and McGuire, P. M., Histone messenger RNA from HeLa cells: evidence for modificated 5' termini, *Biochemistry*, 16, 2207, 1977.

249. Berk, A. J. and Sharp, P. A., Spliced early mRNAs of simian virus 40, *Proc. Natl. Acad. Sci. U.S.A.*, 75, 1274, 1978.

250. Javaherian, K. and Sadeghi, M., Nonhistone proteins HMG1 and HMG2 unwind the DNA double helix, *Nucl. Acids Res.*, 6, 3569, 1979.

251. Javaherian, K., Liu, L. F., and Wang, J. C., Nonhistone proteins HMG1 and HMG2 change the DNA helical structure, *Science*, 199, 1345, 1979.

252. Javaherian, K. and Amini, S., Conformational study of calf thymus HMG14 nonhistone protein, *Biochem. Biophys. Res. Commun.*, 85, 1385, 1978.

253. Abercrombie, B. D., Kneale, G. G., Crane-Robinson, C., Bradbury, M., Goodwin, G. H., Walder, J., and Johns, E. W., Studies on the conformational properties of HMG17 and its interaction with DNA, *Eur. J. Biochem.*, 84, 173, 1978.

254. Jackson, P. J., Fishback, J. L., Bidney, D. L., and Reeck, G. R., Preferential affinity of the high molecular weight HMG nonhistone chromosomal proteins for single stranded DNA, *J. Biol. Chem.*, 254, 5569, 1979.

255. Isackson, J., Clow, L. G., and Reeck, G. R., Comparison of the salt dissociations of the high molecular weight HMG nonhistone chromosomal proteins from double stranded DNA and from chromatin, *FEBS Lett.*, 125, 30, 1981.

256. Shelton, K. R. and Allfrey, V. G., Selective synthesis of a nuclear acidic protein in liver cells stimulated by cortisol, *Nature (London)*, 228, 132, 1980.

257. Enea, V. and Allfrey, V. G., Selective synthesis of liver nuclear acidic proteins following glucagon administration *in vivo*, *Nature (London)*, 242, 265, 1973.

258. Helmsing, P. and Berendes, H., Induced accumulation of nonhistone proteins in polytene nuclei of *Drosophila hydei*, *J. Cell Biol.*, 50, 893, 1971.

259. Krebs, E. G., Protein kinases, *Curr. Top. Cell. Regul.*, 5, 99, 1972.

260. Segal, H. L., Enzymatic interconversion of active and inactive forms of enzymes, *Science*, 180, 25, 1973.

261. Greengard, P., Phosphorylated proteins as physiological effectors, *Science*, 199, 146, 1978.

262. Hughes, M. R., Compton, J. G., Schrader, W. T., and O'Malley, B. W., Interaction of the chick oviduct progesterone receptor with deoxyribonucleic acid, *Biochemistry*, 20, 2481, 1981.

263. Barrack, E. R. and Coffey, D. S., The specific binding of estrogens and androgens to the nuclear matrix of sex hormone responsive tissue, *J. Biol. Chem.*, 255, 7265, 1980.

264. Smith, P. and Von Holt, C., Interaction of the activated cytoplasmic glucocorticoid receptor complex with the nuclear envelope, *Biochemistry*, 20, 2900, 1981.

265. **Kaufmann, S. H., Coffey, D. S., and Shaper, J. H.,** Considerations in the isolation of rat liver nuclear matrix, nuclear envelope and pore complex lamina, *Exp. Cell Res.,* 132, 105, 1981.
266. **Berezney, R. and Coffey, D. S.,** Identification of a nuclear protein matrix, *Biochem. Biophys. Res. Commun.,* 60, 1410, 1974.
267. **Berezney, R. and Coffey, D. S.,** The nuclear protein matrix: isolation, structure and function, *Adv. Enzyme Regul.,* 14, 63, 1976.
268. **Berezney, R.,** Dynamic properties of the nuclear matrix, in *The Cell Nucleus,* Vol. 8, Busch, H., Ed., Academic Press, New York, 1979, 413.
269. **Peters, K. E. and Comings, D. E.,** Two-dimensional gel electrophoresis of rat liver nuclear washes, nuclear matrix, and hnRNA proteins, *J. Cell Biol.,* 86, 135, 1980.
270. **Fakan, S., Puvion, E., and Spohr, G.,** Localization and characterization of newly synthesized nuclear RNA in isolated rat hepatocytes, *Exp. Cell Res.,* 99, 155, 1976.
271. **Herman, R., Weymouth, L., and Penman, S. J.,** Heterogeneous nuclear RNA-protein fibers in chromatin-depleted nuclei, *J. Cell Biol.,* 78, 663, 1978.
272. **Miller, T. E., Huang, L-Y., and Pogo, O. J.,** Rat liver nuclear skeleton and small molecular weight RNA species, *J. Cell Biol.,* 76, 692, 1978.
273. **Milnar, G. R.,** Changes in chromatin structure during interphase in human normoblasts, *Nature New Biol.,* 221, 71, 1969.
274. **Jackson, D. A., McCready, S. J., and Cook, P. R.,** RNA is synthesized at the nuclear cage, *Nature (London),* 292, 552, 1981.
275. **Gilbert, W. and Mueller-Hill, B.,** Isolation of the *lac* repressor, *Proc. Natl. Acad. Sci. U.S.A.,* 56, 1891, 1966.
276. **Adler, K., Beyreuther, K., Fanning, E., Geisler, N., Gronenborn, B., Klemm, A., Mueller-Hill, B., Pfahl, M., and Schmitz, A.,** How *lac* repressor binds to DNA, *Nature (London),* 237, 322, 1972.
277. **Zubay, G., Schwartz, D., and Beckwith, J.,** Mechanism of activation of catabolite-sensitive genes: a positive control system, *Proc. Natl. Acad. Sci. U.S.A.,* 66, 104, 1970.
278. **Anderson, W. B., Schneider, A. B., Emmer, M., Perlman, P. L., and Pastan, I.,** Purification of and properties of the cyclic adenosine 3',5'-monophosphate receptor protein which mediates cyclic adenosine 3',5'-monophosphate-dependent gene transcription in *Escherichia coli, J. Biol. Chem.,* 246, 5929, 1971.
279. **Jagodzinski, L. L., Chilton, J. C., and Sevall, J. S.,** DNA-binding nonhistone proteins: DNA site reassociation, *Nucl. Acids Res.,* 5, 1487, 1978.
280. **Jagodzinski, L. L., Castro, E., Sherrod, P., Lee, D., and Sevall, J. S.,** Reassociation kinetics of nonhistone-bound DNA-sites, *J. Biol. Chem.,* 254, 3038, 1979.
281. **Hsieh, T. S. and Brutlag, D. L.,** A protein that preferentially binds *Drosophila* DNA, *Proc. Natl. Acad. Sci. U.S.A.,* 76, 726, 1979.
282. **Weideli, H., Brack, C., and Gehring, W. J.,** Characterization of *Drosophila* DNA binding protein DB-Z: demonstration of its sequence-specific interaction with DNA, *Proc. Natl. Acad. Sci. U.S.A.,* 77, 3773, 1980.
283. **Dastugue, B. and Crepin, M.,** Interaction of nonhistone proteins with DNA-bound nonhistone proteins, *Eur. J. Biochem.,* 99, 499, 1979.
284. **Crepin, M. and Dastugue, B.,** Regulation of transcription by DNA-bound nonhistone proteins, *Eur. J. Biochem.,* 99, 499, 1979.
285. **Chao, M. N., Gralla, J. D., and Martinson, H. G.,** Lac operator nucleosomes. I. Repressor binds specifically to operator with the nucleosome core, *Biochemistry,* 19, 3254, 1980.
286. **Chao, M. N., Gralla, J. D., and Martinson, H. G.,** Lac operator nucleosomes. II. Lac nucleosomes can change conformation to strengthen binding by lac repressor, *Biochemistry,* 19, 3260, 1980.
287. **Gates, D. M. and Bekhor, I.,** DNA sequences selection by tightly bound nonhistone chromosomal proteins, *Nucl. Acids Res.,* 6, 1617, 1979.
288. **Bekhor, I. and Mirrell, C. J.,** Simple isolation of DNA hydrophobically complexed with presumed gene regulatory proteins (M3), *Biochemistry,* 18, 609, 1979.
289. **Bekhor, I. and Feldman, G.,** Assembly of DNA with histone and nonhistone proteins *in vitro, Biochemistry,* 15, 4771, 1976.
290. **Gates, D. M. and Bekhor, I.,** DNA-binding activity of tightly bound nonhistone chromosomal proteins in chicken liver chromatin, *Nucl. Acid. Res.,* 6, 3411, 1979.
291. **Norman, G. L. and Bekhor, I.,** Enrichment of selected active human gene sequences in the placental DNA fraction associated with tightly bound nonhistone chromosomal proteins, *Biochemistry,* 20, 3568, 1981.
292. **Bowen, B., Steinberg, J., Laemmli, U. K., and Weintraub, H.,** The detection of DNA-binding proteins by blotting, *Nucl. Acid Res.,* 8, 1, 1980.
293. **Burnette, W. N.,** 'Western Blotting': electrophoretic transfer of proteins from sodium dodecyl sulfate-polyacrylamide gels to unmodified nitrocellulose and radiographic detection with antibody and radioiodinated protein A, *Anal. Biochem.,* 112, 195, 1981.

294. **Alberts, B. M. and Frey, L.,** T4 bacteriophage gene 32: a structural protein in the replication and recombination of DNA, *Nature (London),* 227, 1313, 1970.

295. **Scherzinger, E., Hitfin, F., and Jost, E.,** Stimulation of T7 DNA polymerase by a new phage-coded protein, *Mol. Gen. Genet.,* 123, 247, 1973.

296. **Otto, B., Baynes, M., and Knippers, R.,** A single stranded specific DNA-binding protein from mouse cells that stimulates DNA polymerase; it's modification by phosphorylation, *Eur. J. Biochem.,* 73, 17, 1977.

297. **Carrara, G., Gattoni, S., Mercanst, B., and Tochini-Valentini, G. P.,** Purification of a DNA-binding protein from *Xenopus laevis* unfertilized eggs, *Nucl. Acids Res.,* 4, 2855, 1977.

298. **Kohwi-Shigematsu, I., Enomoto, T., Yamada, M.-A., and Nakanishi, M.,** Exposure of DNA bases induced by the interactions of DNA and calf thymus DNA helix destabilizing protein, *Proc. Natl. Acad. Sci. U.S.A.,* 75, 4689, 1978.

299. **Durnford, J. M. and Champoux, J. J.,** The DNA untwisting enzyme from *Sacharomyces cerevisiae,* *J. Biol. Chem.,* 253, 1086, 1981.

300. **Duguet, M. and de Recondo, A. M.,** A DNA unwinding protein isolated from regenerating rat liver, *J. Biol. Chem.,* 253, 1660, 1978.

301. **Mather, J. and Hotta, Y.,** A phosphoratable DNA-binding protein associated with a lipoprotein fraction from rat spermatocyte nuclei, *Exp. Cell Res.,* 109, 181, 1977.

302. **Hotta, Y. and Stern, H.,** The effect of dephosphorylation on the properties of a helix destabilizing protein from meiotic cells and its partial reversal by protein kinase, *Eur. J. Biochem.,* 95, 31, 1979.

303. **Nass, K. and Frenkel, G. D.,** A DNA-binding protein from KB cells which inhibits deoxyribonuclease activity on single strand DNA, *J. Biol. Chem.,* 254, 3407, 1979.

304. **Krajewska, W., Harbec, E., and Klyszejko-Stephanowicz, L.,** Changes in DNA-binding chromosomal nonhistone proteins during chicken erythroid cell maturation, *Biochimie,* 60, 211, 1978.

305. **Yeoman, L. C., Taylor, C. W., Jordan, C. W., and Busch, H.,** Differences in chromatin proteins of growing and non-growing tissues, *Exp. Cell Res.,* 91, 207, 1975.

306. **Magim, B. E. and Dorsett, P. H.,** Changes in the synthesis of DNA-binding proteins during the onset of transformation in NRK cells transfomed by TS mutant of Rous Sarcoma virus, *J. Virol.,* 22, 469, 1977.

307. **Spelsberg, T. C.,** *Acidic Proteins of the Nucleus,* Cameron, I. L. and Jeter, J. R., Eds., Academic Press, New York, 1974, 247.

308. **Chiu, J. F., Fujitani, H., and Hnilica, L. S.,** Methods for selective extraction of chromosomal nonhistone proteins, in *Methods in Cell Biology,* Vol. 16, Academic Press, New York, 1977, 286.

309. **Busch, H., Ballal, R., Busch, R. K., Ezrailson, E., Goldknopf, I. L., Olson, M. O. J., Prestayko, A. W., Taylor, C. W., and Yeoman, L. C.,** Chromatin Proteins: electrophoretic, immunologic and metabolic characteristics, in *Chromosomal Proteins and Their Role in the Regulation of Gene Expression,* Stein, G. S., Stein, J. L., and Kleinsmith, L. J., Eds., Academic Press, New York, 1975, 187.

310. **Gilmour, R. S. and Paul, J.,** Role of non-histone components in determining organ specificity of rabbit chromatins, *FEBS Lett.,* 9, 242, 1970.

311. **Paul, J. and Gilmour, R. S.,** Organ-specific restriction of transcription in mammalian chromatin, *J. Mol. Biol.,* 34, 305, 1968.

312. **Gilmour, R. S. and Paul, J.,** RNA transcribed from reconstituted nucleoprotein is similar to natural RNA, *J. Mol. Biol.,* 40, 137, 1969.

313. **Stein, G. S., Stein, J. L., Baumbach, L., Leza, A., Lichtler, A., Marashi, F., Plumb, M., Rickles, R., Sierra, F., and Van Dyke, T.,** Regulation of Histone Gene Expression in Human Cells, Miami Winter Symposia, Vol. 16, in press.

314. **Wang, T. Y.,** Activation of transcription *in vitro* from chromatin by nonhistone proteins, *Exp. Cell Res.,* 61, 455, 1970.

315. **Kamiyama, M. and Wang, T. Y.,** Activated transcription from rat liver chromatin by non-histone proteins, *Biochim. Biophys. Acta,* 228, 563, 1971.

316. **Kostraba, N. C. and Wang, T. Y.,** Differential activation of transcription of chromatin by nonhistone proteins, *Biochim. Biophys. Acta,* 262, 169, 1972.

317. **Stein, G. S. and Farber, J. L.,** The role of nonhistone chromosomal proteins in the restriction of mitotic chromatin template activity, *Proc. Natl. Acad. Sci. U.S.A.,* 69, 2918, 1972.

318. **Stein, G. S., Chaudhuri, S. K., and Baserga, R.,** Gene activation in WI-38 fibroblasts stimulated to proliferate: role of nonhistone chromosomal proteins, *J. Biol. Chem.,* 247, 3019, 1972.

319. **Rickwood, D., Threlfall, G., MacGillivray, A. J., Paul, J., and Riches, P.,** Studies on the phosphorylation of chromatin non-histone proteins and their effect on deoxyribonucleic acid transcription, *Biochem. J.,* 129, 50, 1972.

320. **Shea, M. and Kleinsmith, L. J.,** Template-specific stimulation of RNA synthesis by phosphorylated non-histone chromatin proteins, *Biochem. Biophys. Res. Commun.,* 50, 473, 1973.

321. **Peterson, J. L. and McConkey, E. H.,** Proteins of Friend leukemia cells. Comparison of hemoglobin-synthesizing and noninduced populations, *J. Biol. Chem.,* 251, 555, 1976.

322. Salem, R. and DeVellis, J., Protein kinase activity and cAMP-dependent protein phosphorylation in subcellular fractions after norepinephrine treatment of glial cells, *Fed. Proc.*, 35, 296, 1976.

323. Kamiyama, M., Dastugue, B., Defer, N., and Kruh, J., Liver chromatin non-histone proteins. Partial fractionation and mechanism of action on RNA synthesis, *Biochim. Biophys. Acta*, 277, 576, 1972.

324. Kostraba, N. C., Montagna, R. A., and Wang, T. Y., Study of the loosely bound non-histone chromatin proteins. Stimulation of deoxyribonucleic acid-binding phosphoprotein fraction, *J. Biol. Chem.*, 250, 1548, 1975.

325. Kostraba, N. C. and Wang, T. Y., Inhibition of transcription *in vitro* by a non-histone protein isolated from Ehrlich ascites tumor chromatin, *J. Biol. Chem.*, 250, 8938, 1975.

326. Kostraba, N. C., Montagna, R. A., and Wang, T. Y., Mode of action of nonhistone proteins in the stimulation of transcription from DNA, *Biochem. Biophys. Res. Commun.*, 72, 334, 1976.

327. Kostraba, N. C., Newman, R. S., and Wang, T. Y., Selective inhibition of transcription by a non-histone protein isolated from Ehrlich ascites tumor chromatin, *Arch. Biochem. Biophys.*, 179, 100, 1977.

328. Seifart, K. H., A factor stimulating the transcription on double-stranded DNA by purified RNA polymerase from rat liver nuclei, *Cold Spring Harbor Symp. Quant. Biol.*, 35, 719, 1970.

329. Stein, H. and Hausen, F., A factor from calf thymus stimulating DNA-dependent RNA polymerase isolated from this tissue, *Eur. J. Biochem.*, 14, 270, 1970.

330. DiMauro, E., Hollenberg, C. P., and Hall, B. D., Transcription in yeast: a factor that stimulates yeast RNA polymerases, *Proc. Natl. Acad. Sci. U.S.A.*, 69, 2818, 1972.

331. Lee, S.-C. and Dahmus, M. E., Stimulation of eukaryotic DNA-dependent RNA polymerase by protein factors, *Proc. Natl. Acad. Sci. U.S.A.*, 70, 1383, 1973.

332. Sekimizu, K., Kobayashi, N., Mizuno, D., and Natori, S., Purification of a factor from Ehrlich ascites tumor cells specifically stimulating RNA polymerase II, *Biochemistry*, 15, 5064, 1976.

333. James, G. T., Yeoman, L. C., Matsui, S., Goldberg, A. H., and Busch, H., Isolation and characterization of nonhistone chromosomal protein C-14 which stimulates RNA synthesis, *Biochemistry*, 16, 2384, 1977.

334. Higashinakagawa, T., Onishi, T., and Muramatsu, M., A factor stimulating the transcription by nucleolar RNA polymerase in the nucleolus of rat liver, *Biochem. Biophys. Res. Commun.*, 48, 937, 1972.

335. Kuroiwa, A., Mizuno, D., and Natori, S., Protein which interacts with a stimulatory factor of RNA polymerase II of Ehrlich ascites tumor cells, *Biochemistry*, 16, 5687, 1977.

336. O'Malley, B. W., Towle, H. C., and Schwartz, R. J., Regulation of gene expression in eucaryotes, *Annu. Rev. Genet.*, 11, 239, 1977.

337. Paul, J., Gilmour, R. S., Affara, N., Birnie, G., Harrison, P., Hell, A., Humphries, S., Windass, J., and Young, B., The globin gene: structure and expression, *Cold Spring Harbor Symp. Quant. Biol.*, 38, 885, 1973.

338. Barrett, T., Maryanka, D., Hamlyn, P., and Gould, H. J., Nonhistone proteins control gene expression in reconstituted chromatin, *Proc. Natl. Acad. Sci. U.S.A.*, 71, 5057, 1974.

339. Chiu, J. F., Tsai, Y. H., Sakuma, K., and Hnilica, L. S., Regulation of *in vitro* mRNA transcription by a fraction of chromosomal proteins, *J. Biol. Chem.*, 250, 9431, 1975.

340. Tsai, S. Y., Harris, S. E., Tsai, M. J., and O'Malley, B. W., Effects of estrogen on gene expression in chick oviduct. The role of chromatin proteins in regulating transcription of the ovalbumin gene, *J. Biol. Chem.*, 251, 4713, 1976.

341. Borun, T. W., Scharf, M. D., and Robbins, E., Rapidly labeled, polyribosome-associated RNA having the properties of histone messenger, *Proc. Natl. Acad. Sci. U.S.A.*, 58, 1977, 1967.

342. Jacobs-Lorena, M., Baglioni, C., and Borun, T. W., Translation of messenger RNA for histones from HeLa cells by a cell-free extract from mouse ascites tumor, *Proc. Natl. Acad. Sci. U.S.A.*, 69, 2095, 1972.

343. Gallwitz, D. and Breindl, M., Synthesis of histones in a rabbit reticulocyte cell-free system directed by a polyribosomal RNA fraction from synchronized HeLa cells, *Biochem. Biophys. Res. Commun.*, 47, 1106, 1972.

344. Stein, J. L., Thrall, C. L., Park, W. D., Mans, R. J., and Stein, G. S., Hybridization analysis of histone messenger RNA association with polyribosomes during the cell cycle, *Science*, 189, 557, 1975.

345. Detke, S., Lichtler, A., Philips, I., Stein, J., and Stein, G., Reassessment of histone gene expression during cell cycle in human cells by using homologous H4 histone cDNA, *Proc. Natl. Acad. Sci. U.S.A.*, 76, 1995, 1979.

346. Stein, G. S., Park, W. D., Thrall, C. L., Mans, R. J., and Stein, J. L., Regulation of histone gene transcription during the cell cycle by nonhistone chromosomal proteins, *Nature (London)*, 257, 764, 1975.

347. Stein, G. S., Stein, J. L., Park, W. D., Detke, S., Lichtler, A. C., Shephard, E. A., Jansing, R. L. and Phillips, I. R., Regulation of gene expression in HeLa S₃ cells, *Cold Spring Harbor Symp. Quant. Biol.*, 42, 1107, 1977.

348. Detke, S., Stein, J., and Stein, G., Synthesis of histone messenger RNAs by RNA polymerase II in nuclei from S phase HeLa S3 cells, *Nucl. Acids Res.*, 5, 1515, 1978.

349. Jansing, R. L., Stein, J. L., and Stein, G. S., Activation of histone gene transcription by nonhistone chromosomal proteins in WI-38 human diploid fibroblasts, *Proc. Natl. Acad. Sci. U.S.A.*, 74, 173, 1977.

350. Melli, M., Spinelli, G., and Arnold, E., Synthesis of histone messenger RNA of HeLa cells during the cell cycle, *Cell*, 12, 167, 1977.

351. Stein, J. L., Stein, G. S., and McGuire, P. M., Histone mRNA from the HeLa cells: evidence for modified 5′ termini, *Biochemistry*, 16, 2207, 1977.

352. Stein, G. S. and Borun, T. W., The synthesis of acidic chromosomal proteins during the cell cycle of HeLa S3 cells. I. The accelerated accumulation of acidic residual nuclear protein before the initiation of DNA replication, *J. Cell Biol.*, 52, 292, 1972.

353. Baserga, R., Non-histone chromosomal proteins in normal and abnormal growth, *Life Sci.*, 15, 1057, 1974.

354. Stein, G. S. and Farber, J. L., The role of nonhistone chromosomal proteins in the restriction of mitotic chromatin templates activity, *Proc. Natl. Acad. Sci. U.S.A.*, 69, 2918, 1972.

355. Stein, G. S., Chaudhuri, S. K., and Baserga, R., Gene activation in WI-38 fibroblasts stimulated to proliferate: role of nonhistone chromosomal proteins, *J. Biol. Chem.*, 247, 3019, 1972.

356. Bekhor, I., Kung, G. M., and Bonner, J., Sequence-specific interaction of DNA and chromosomal proteins, *J. Mol. Biol.*, 39, 351, 1969.

357. Paul, J., Gilmour, R. S., Affara, N., Birnie, G., Harrison, P., Hell, A., Humphries, S., Windass, J., and Young, B., The globin gene: structure and expression, *Cold Spring Harbor Symp. Quant. Biol.*, 38, 885, 1973.

358. Barrett, T., Maryanka, D., Hamlyn, P., and Gould, H. J., Nonhistone proteins control gene expression in reconstituted chromatin, *Proc. Natl. Acad. Sci. U.S.A.*, 71, 5057, 1974.

359. Chiu, J. F., Tsai, Y. H., Sakuma, K., and Hnilica, L. S., Regulation of *in vitro* mRNA transcription by a fraction of chromosomal proteins, *J. Biol. Chem.*, 250, 9431, 1975.

360. Yu, L. C., Szabo, P., Borun, T. W., and Prensky, W., The localization of the genes coding for histone H4 in human chromosomes, *Cold Spring Harbor Symp. Quant. Biol.*, 42, 1101, 1978.

361. Chandler, M. E., Kedes, L. H., Cohen, R. H., and Yunis, J. J., Genes coding for histone proteins in man are located on the distal end of the long arm of chromosome 7, *Science*, 205, 908, 1979.

362. Naylor, S., Stein, J. L., and Stein, G. S., unpublished results.

363. Stein, G. S., Stein, J. L., Laipis, P. J., Chattopadhyay, S. K., Lichtler, A. C., Detke, S., Thomson, J. A., Phillips, I. R., and Shephard, E. A., *Regulation of Gene Expression in Normal and Neoplastic Cells*, Miami Winter Symposia, Vol. 15, Academic Press, New York, 1981, 125.

364. Stein, G. S., Stein, J. L., Park, W. D., Detke, S., Lichtler, A. C., Shephard, E. A., Jansing R. H., and Phillips, I. R., Regulation of histone gene transcription in HeLa S3 cells, *Cold Spring Harbor Symp. Quant. Biol.*, 42, 1107, 1977.

365. Lichtler, A. C., Detke, S., Phillips, I. R., Stein, G. S., and Stein, J. L., Multiple forms of H4 histone mRNA in human cells, *Proc. Natl. Acad. Sci. U.S.A.*, 77, 1942, 1980.

366. Grunstein, M., Levy, S., Schedl, P., and Kedes, L., Messenger RNAs for individual histone proteins: fingerprint analysis and *in vitro* translation, *Cold Spring Harbor Symp. Quant. Biol.*, 38, 717, 1973.

367. Berger, S. L. and Birkenmeier, C. S., Inhibition of intractable nucleases with ribonucleoside vanadyl complexes: isolation of messenger ribonucleic acid from resting lymphocytes, *Biochemistry*, 18, 5143, 1979.

368. Peattie, D. A., Direct chemical method for sequencing RNA, *Proc. Natl. Acad. Sci. U.S.A.*, 76, 1760, 1979.

369. Volckaert, G., Minjon, W., and Fiers, W., Analysis of ^{32}P-labeled bacteriophage MS3 RNA by a mini-fingerprinting procedure, *Ann. Biochem.*, 72, 433, 1976.

370. Alwine, J. L., Kemp, D. J., Parker, B. A., Reiser, J., Renart, J., Stark, G. R., and Wahl, G. M., Detection of specific RNAs or specific fragments in gels and transfer to diazobenzyloxymethyl paper, *Methods Enzymol.*, 68, 220, 1979.

371. Stellwag, E. J. and Dahlberg, A. E., Electrophoretic transfer of DNA, RNA and protein onto diazobenzyloxymethyl (DBM)-paper, *Nucl. Acids Res.*, 8, 299, 1980.

372. Haegeman, G., Iserentant, D., Gheysen, D., and Fiers, W., Characterization of the major altered leader sequence of late mRNA induced by SV40 deletion mutant d1-1811, *Nucl. Acids Res.*, 7, 1799, 1979.

373. Rickles, R., Marashi, F., Sierra, F., Clark, S., Wells, J., Stein, J., and Stein, G. S., Analysis of histone gene expression during the cell cycle in HeLa cells using cloned human histone genes, *Proc. Natl. Acad. Sci. U.S.A.*, 79, 749, 1982.

374. Honda, B. M. and Roeder, R. G., Association of a 5S gene transcription factor with 5S RNA and altered levels of the factor during cell differentiation, *Cell*, 22, 119, 1980.

375. Weintraub, H. and Groudine, M., Chromosomal subunits in active genes have an altered conformation, *Science*, 193, 848, 1976.

376. Flint, S. J. and Weintraub, H. M., An altered subunit configuration associated with the actively transcribed DNA of integrated adenovirus genes, *Cell*, 12, 783, 1977.

377. Lohr, D. and Hereford, L., Yeast chromatin is uniformly digested by DNase 1, *Proc. Natl. Acad. Sci. U.S.A.*, 76, 4285, 1979.

378. Garel, A., Zolan, M., and Axel, R., Genes transcribed at diverse rates have a similar conformation in chromatin, *Proc. Natl. Acad. Sci. U.S.A.*, 74, 486, 1979.

379. Levy, W. B. and Dixon, G. H., Renaturation kinetics of cDNA complementary to cytoplasmic polyadenylated RNA from rainbow trout testis. Accessibility of transcribed genes to pancreatic DNase, *Nucl. Acids Res.*, 4, 883, 1977.

380. Miller, D. M., Turner, P., Nieuhuis, A. W., Axelrod, D. E., and Gopalakrishnan, T. V., Active conformation of the globin genes in uninduced and induced mouse erythroleukemia cells, *Cell*, 14, 511, 1978.

381. Bloom, K. S. and Anderson, J. N., Conformation of ovalbumin and globin genes in chromatin during differential gene expression, *J. Biol. Chem.*, 254, 10532, 1979.

382. Weintraub, H., Larsen, A., and Groudine, M., α-Globin-gene switching during the development of chicken embryos: expression and chromatin structure, *Cell*, 24, 333, 1981.

383. Groudine, M. and Weintraub, H., Activation of globin genes during chicken development, *Cell*, 24, 393, 1981.

384. Febber, B. K., Gerber-Huber, S., Meier, C., May, F. E. B., Westley, B., Weber, R., and Ryffel, G. U., Quantitation of DNase 1 sensitivity in *Xenopus* chromatin containing active and inactive globin, albumin and vitellogenin genes, *Nucl. Acids Res.*, 9, 2455, 1981.

385. Gerber-Huber, S., Febber, B. K., Weber, R., and Ryffel, G. U., Estrogen induces tissue specific changes in the chromatin conformation of the vitellogenin genes in *Xenopus*, *Nucl. Acids Res.*, 9, 2475, 1981.

386. Stadler, J., Larsen, A., Engel, J. D., Dolan, M., Groudine, M., and Weintraub, H., Tissue specific DNA cleavages in the globin chromatin domain introduced by DNase 1, *Cell*, 20, 451, 1980.

387. Giri, C. P. and Gorovsky, M. A., DNase 1 sensitivity of ribosomal genes in isolated nucleosome core particles, *Nucl. Acids Res.*, 5, 197, 1980.

388. Senear, A. W. and Palmiter, R. D., Multiple structural features are responsible for the nuclease sensitivity of the active ovalbumin gene, *J. Biol. Chem.*, 256, 1191, 1981.

389. Weisbrod, S., Groudine, M., and Weintraub, H., Interaction of HMG 14 and HMG 17 and an analysis of a globin chromatin, *Cell*, 11, 389, 1981.

390. Weisbrod, S. and Weintraub, H., Isolation of actively transcribed nucleosomes using immobilized HMG 14 and HMG 17 and an analysis of α-globin chromatin, *Cell*, 23, 391, 1981.

391. Wu, C. and Gilbert, W., Tissue-specific exposure of chromatin structure at the 5′ terminus of the rat pre-proinsulin II gene, *Proc. Natl. Acad. Sci. U.S.A.*, 78, 1577, 1981.

392. Keene, M. A., Corces, V., Lowenhaupt, K. Y., and Elgin, S. C. R., DNase 1 hypersensitive sites in Drosophila chromatin occur at 5′ ends of regions of transcription, *Proc. Natl. Acad. Sci. U.S.A.*, 78, 143, 1981.

393. Samal, B., Worcel, A., Louis, C., and Schedl, P., Chromatin structure of the histone genes of *D. melanogaster*, *Cell*, 23, 401, 1981.

394. Neelin, J. M., Mazen, A., and Champagne, M., The fractionation of active and inactive chromatins from erythroid cells of chicken, *FEBS Lett.*, 65, 309, 1976.

395. Gottesfeld, J. M. and Partington, G. A., Distribution of messenger RNA-coding sequences in fractionated chromatin, *Cell*, 12, 953, 1977.

396. Lau, A. F., Ruddon, R. W., Collett, M. S., and Faras, A. J., Distribution of the globin gene in active and inactive chromatin fractions from Friend erythroleukemia cells, *Exp. Cell Res.*, 111, 269, 1978.

397. Matusi, S. and Busch, H., Isolation and characterization of rDNA-containing chromatin from nucleoli, *Exp. Cell Res.*, 109, 151, 1977.

398. Alberga, A., Trau, A., and Baulieu, E. E., Distribution of estradiol receptor and vitellogenin gene in chick liver chromatin fractions, *Nucl. Acid Res.*, 7, 2031, 1979.

399. Levy, W. B. and Dixon, G. H., Partial purification of transcriptionally active nucleosomes from trout testes cells, *Nucl. Acid Res.*, 5, 4155, 1978.

400. Bloom, K. S. and Anderson, J. N., Fractionation of hen oviduct chromatin into transcriptionally active and inactive regions after selective micrococcal nuclease digestion, *Cell*, 15, 141, 1978.

401. Reeves, R. and Jones, A., Genomic transcriptional activity and the structure of chromatin, *Nature (London)*, 260, 495, 1976.

402. Garel, A. and Axel, A., The structure of transcriptionally active ovalbumin genes in chromatin, *Cold Spring Harbor Symp. Quant. Biol.*, 42, 701, 1977.

403. Louis, C., Schedl, P., Samal, B., and Worcel, A., Chromatin structure of the 5s DNA genes in *D. melanogaster*, *Cell*, 22, 387, 1980.

404. **Levy, W. B., Connor, W., and Dixon, G. H.,** A subset of trout testis nucleosomes enriched in transcribed DNA sequences containing HMG proteins as major structural components, *J. Biol. Chem.,* 254, 609, 1979.

405. **Olins, A. L., Carlson, R. D., Wright, E. B., and Olins, D. E.,** Chromatin γ bodies: isolation, subfractionation and physical characterization, *Nucl. Acids Res.,* 3, 3271, 1976.

406. **Bourgoyne, L. A. and Skinner, J. D.,** Chromatin superstructure: the next level of structure above the nucleosome has an alternating character. A two-nucleosome based series is generated by probes armed with DNAse-1 acting on isolated nuclei, *Biochem. Biophys. Res. Commun.,* 99, 893, 1981.

407. **Kachatrian, A. T., Pospelov, V. A., Svetlikova, S. B., and Vorob'ev, V. I.,** Nucleodisome — a new repeat of chromatin revealed in nuclei of pigeon erythrocytes by DNase 1 digestion, *FEBS Lett.,* 128, 90, 1981.

408. **Riggs, M. G., Whittacker, R. G., Neuman, J. R., and Ingram, V. M.,** n-Butyrate causes histone modification in HeLa and Friend erythroleukemia cells, *Nature (London),* 268, 462, 1977.

409. **Ruiz-Carrillo, A., Wangh, L. J., and Allfrey, V. G.,** Processing of newly synthesized histone molecules, *Science,* 190, 117, 1975.

410. **Adler, A. J., Fasman, G. D., Wangh, L. J., and Allfrey, V. G.,** Altered conformational effects of naturally acetylated F2$_a$(lv) in F2$_a$1-deoxyribonucleic acid complexes, *J. Biol. Chem.,* 249, 2911, 1974.

411. **Allfrey, V. G.,** Structural modification of histones and their possible role in the regulation of ribonucleic acid synthesis, *Can. Cancer Conf.,* 6, 313, 1964.

412. **Allfrey, V. G., Faulkner, R., and Mirsky, A. E.,** Acetylation and methylation of histone and their possible role in the regulation of RNA synthesis, *Proc. Natl. Acad. Sci. U.S.A.,* 51, 786, 1964.

413. **Dobson, M. E. and Ingram, V. M.,** In vitro transcription of chromatin containing histones hyperacetylated in vivo, *Nucl. Acids Res.,* 5, 4201, 1980.

414. **Pogo, B. G. T., Pogo, A. O., Allfrey, V. G., and Mirsky, A. E.,** RNA synthesis and histone acetylation during the course of gene activation in lymphocytes, *Proc. Natl. Acad. Sci. U.S.A.,* 55, 805, 1966.

415. **Graaff, G. deV. and Von Holt, C.,** Enzymatic histone modification during the induction of tyrosine aminotransferase with insulin and hydrocortisone, *Biochim. Biophys. Acta,* 299, 480, 1973.

416. **Horgen, P. A. and Ball, S. F.,** Nuclear protein acetylation during hormone-induced sexual differentiation in *Achlya ambisexualis, Cytobios,* 10, 181, 1974.

417. **Levy, W. B., Watson, D. C., and Dixon, G. H.,** Multiacetylated forms of H4 are found in a putative transcriptionally competent chromatin fraction from trout testis, *Nucl. Acids Res.,* 6, 259, 1979.

418. **Nelson, D., Covault, J., and Chalkley, R.,** Segregation of rapid acetylated histones into a chromatin fraction released from intact nuclei by the action of micrococcal nuclease, *Nucl. Acids Res.,* 8, 1745, 1980.

419. **Perry, M. and Chalkley, R.,** The effect of histone hyperacetylation on the nuclease sensitivity and the solubility of chromatin, *J. Biol. Chem.,* 256, 3313, 1980.

420. **Vidali, G., Boffa, L. C., Bradbury, E. M., and Allfrey, V. G.,** Butyrate suppression of histone deacetylation leads to accumulation of multiacetylated forms of histones H3 and H4 and increased DNase 1 sensitivity of the associated DNA sequence, *Proc. Natl. Acad. Sci. U.S.A.,* 75, 2239, 1978.

421. **Nelson, D. A., Perry, W. M., and Chalkley, R.,** Sensitivity of regions of chromatin containing hyperacetylated histone proteins to DNase 1, *Biochem. Biophys. Res. Commun.,* 82, 356, 1978.

422. **Bode, J., Henco, K., and Wingender, E.,** Modulation of the nucleosome structure by histone acetylation, *Eur. J. Biochem.,* 110, 143, 1980.

423. **Simpson, R. T.,** Structure of chromatin containing extensively acetylated H3 and H4, *Cell,* 13, 691, 1978.

424. **Blau, H. M. and Epstein, C. J.,** Manipulation of myogenisis in vitro: reversible inhibition by DMSO, *Cell,* 17, 95, 1979.

425. **Samuels, H. H., Stanley, F., Casanova, J., and Shao, T. C.,** Thyroid hormone nuclear receptor levels are influenced by the acetylation of chromatin-associated proteins, *J. Biol. Chem.,* 255, 2499, 1980.

426. **Leder, A. and Leder, P.,** Butyric acid, a potent inducer of erythroid differentiation in culture, *Cell,* 5, 319, 1975.

427. **Leavitt, J. and Moyzis, R.,** Changes in gene expression accompanying neoplastic transformation of Syrian Hamster cells, *J. Biol. Chem.,* 253, 2497, 1978.

428. **McKnight, S. G., Hager, L., and Palmiter, R. D.,** Butyrate and related inhibitors of histone deacetylation block the induction of egg white genes by steroid hormones, *Cell,* 22, 469, 1981.

429. **Lindahl, T.,** DNA methylation and control of gene expression, *Nature (London),* 290, 363, 1981.

430. **Goodwin, G. H. and Johns, E. W.,** Isolation and characterization of two calf thymus chromatin nonhistone proteins with high contents of acidic and basic amino acids, *Eur. J. Biochem.,* 40, 215, 1973.

431. **Sanders, C.,** A method for the fractionation of the HMG nonhistone chromosomal proteins, *Biochem. Biophys. Res. Commun.,* 78, 1034, 1977.

432. Watson, D. C., Wong, N. C., and Dixon, G. H., The complete amino acid sequence of a trout testis nonhistone protein H6 localized in a subset of nucleosomes and its similarity to calf thymus non-histone proteins HMG 14 and HMG 17, *Eur. J. Biochem.*, 95, 193, 1979.

433. Rabbani, A., Goodwin, G. H., and Johns, E. W., HMG nonhistone chromosomal proteins from chicken erythrocytes, *Biochem. Biophys. Res. Commun.*, 81, 351, 1978.

434. Mathew, C. G. P., Goodwin, G. H., Gooderham, K., Walker, J. M., and Johns, E. W., A comparison of the HMG nonhistone chromosomal protein in chicken thymus and erythrocytes, *Biochem. Biophys. Res. Commun.*, 87, 1243, 1979.

435. Rabbani, A., Goodwin, G. H., and Johns, E. W., Studies on the tissue specificity of the HMG nonhistone chromosomal proteins from calf, *Biochem. J.*, 173, 497, 1978.

436. Hyde, J. F., Igo-Kemenes, T., and Zachau, H. G., The nonhistone proteins of the rat liver nucleus and their distribution amongst chromatin fractions as produced by nuclease digestion, *Nucl. Acids Res.*, 7, 31, 1979.

437. Levy, W. B., Wong, N. C. W., Watson, D. C., Peters, E. H., and Dixon, G. H., Structure and function of the low-salt extractable chromosomal proteins: preferential association of trout testis proteins H6 and HMGT with chromatin regions selectively sensitive to nucleases, *Cold Spring Harbor Symp. Quant. Biol.*, 42(2), 793, 1977.

438. Mathew, C. G. P., Goodwin, G. H., and Johns, E. W., Studies on the association of the HMG nonhistone proteins with isolated nucleosomes, *Nucl. Acids Res.*, 6, 167, 1979.

439. Levy, W. B. and Dixon, G. H., A study on the localization of the HMGs in chromatin, *Can. J. Biochem.*, 56, 480, 1978.

440. Bakayev, V. V., Bakayeva, G., Schmatenko, V. V., and Georgiev, G. P., Nonhistone proteins in mononucleosomes and subnucleosomes, *Eur. J. Biochem.*, 91, 291, 1978.

441. Goodwin, G. H., Mathew, C. G. P., Wright, C. A., Venkov, C. D., and Johns, E. W., Analysis of the HMG proteins associated with salt soluble nucleosomes, *Nucl. Acid Res.*, 7, 1815, 1979.

442. Jackson, J. B., Pollock, J. M., and Rill, R. L., Chromatin fractionation procedure that yields nucleosomes containing near stoichiometric amounts of HMG nonhistone chromosomal proteins, *Biochemistry*, 18, 3739, 1979.

443. Jackson, J. B. and Rill, R. L., Circular dichroism, thermal denaturation and DNase digestion studies of nucleosomes highly enriched in HMG proteins HMG1 and HMG2, *Biochemistry*, 20, 1042, 1981.

444. Levy, W. B., Enhanced phosphorylation of HMG proteins in nuclease sensitive mononucleosomes from butyrate-treated HeLa cells, *Proc. Natl. Acad. Sci. U.S.A.*, 78, 2189, 1981.

445. Plumb, M. A. and MacGillivray, A. J., The distribution of HMG proteins in chromatin fractions produced by nuclease digestion of pig thymus nuclei, *Biochem. Soc. Trans.*, 9, 143, 1981.

446. Vidali, G., Boffa, L. C., and Allfrey, V. G., Selective release of chromosomal proteins during limited DNase 1 digestion of avian erythrocytes chromatin, *Cell*, 12, 409, 1977.

447. Levy, W. B., Kuehl, L., and Dixon, G. H., The release of HMG H6 and protamine gene sequences upon selective DNase degradation of trout testes chromatin, *Nucl. Acids Res.*, 8, 2859, 1980.

448. Hamana, K. and Zama, M., Selective release of HMG nonhistone proteins during DNase digestion of *Tetrahymena* chromatin at different stages of the cell cycle, *Nucl. Acid Res.*, 8, 5275, 1981.

449. Goodwin, G. H. and Johns, E. W., Are the HMG nonhistone chromosomal proteins associated with "active chromatin"?, *Biochim. Biophys. Acta*, 519, 279, 1978.

450. Weisbrod, S. and Weintraub, H., Isolation of a subclass of nuclear proteins responsible for conferring a DNase 1 sensitive structure on globin chromatin, *Proc. Natl. Acad. Sci. U.S.A.*, 76, 630, 1979.

451. Georgieva, E. I., Pashev, I. G., and Tsanev, R. G., Distribution of HMG and other acid-soluble proteins in fractionated chromatin, *Biochim. Biophys. Acta*, 652, 240, 1981.

452. Kuehl, L., Lynes, T., Dixon, G. H., and Levy, B. W., Distribution of HMG proteins among domains of trout testes chromatin differring in their susceptibility to micrococcal nuclease, *J. Biol. Chem.*, 255, 1090, 1980.

453. Bakayev, V. V., Schmatenko, V. V., and Georgiev, G. P., Subnucleosome particles containing HMG-E and HMG-G originate in transcriptionally active chromatin, *Nucl. Acids Res.*, 7, 1525, 1979.

454. Abercrombie, B. D. A., Kneale, G. G., Crane-Robinson, C., Bradbury, M., Goodwin, G. H., Walker, J., and Johns, E. W., Studies on the conformational properties of HMG 17 and its interaction with DNA, *Eur. J. Biochem.*, 84, 173, 1978.

455. Walker, J. M., Hastings, J. R. B., and Johns, E. W., The partial amino acid sequence of a nonhistone chromosomal protein, *Biochem. Biophys. Res. Commun.*, 73, 72, 1976.

456. Itkes, A. V., Golotov, B. O., Nikolacv, L. G., and Severin, E. S., Clusters of nonhistone chromosomal protein HMG 1 molecules in intact chromatin, *FEBS Lett.*, 118, 63, 1981.

457. Conner, B. J. and Comings, D. E., Isolation of nonhistone chromosomal HMG protein from mouse liver nuclei by hydrophobic chromatography, *J. Biochem. Chem.*, 258, 3283, 1981.

458. Gazit, B., Panet, A., and Cedar, H., Reconstitution of a DNase 1 structure on active genes, *Proc. Natl. Acad. Sci. U.S.A.*, 77, 1787, 1980.

459. Stein, G. S. and Thrall, C. L., Evidence for the presence of non-histone chromosomal proteins in the nucleoplasm of HeLa S3 cells, *FEBS Lett.*, 32, 41, 1973.

460. Rovera, G., Farber, J. L., and Baserga, R., Gene activation in WI-38 fibroblasts stimulated to proliferate: requirement for protein synthesis, *Proc. Natl. Acad. Sci. U.S.A.*, 68, 1725, 1971.

461. Kleinsmith, L. J., Stein, J., and Stein, G., Dephosphorylation of nonhistone proteins specifically alters the pattern of gene transcription in reconstituted chromatin, *Proc. Natl. Acad. Sci. U.S.A.*, 73, 1174, 1976.

462. Thomson, J. A., Stein, J. L., Kleinsmith, L. J., and Stein, G. S., Activation of histone gene transcription by nonhistone chromosomal phosphoproteins, *Science*, 194, 428, 1976.

463. Levy, W. B., Wong, N. C. W., and Dixon, G. H., Selective association of the trout-specific H6 protein with chromatin regions susceptible to DNase 1 and DNase II: possible location of HMG-T in the spacer region between core nucleosomes, *Proc. Natl. Acad. Sci. U.S.A.*, 74, 2810, 1977.

464. Kostraba, N. C., Loor, R. M., and Wang, T. Y., Tissue-specificity of the nonhistone protein that inhibits RNA synthesis *in vitro*, *Biochem. Biophys. Res. Commun.*, 79, 347, 1977.

INDEX

Printed and bound by CPI Group (UK) Ltd, Croydon, CR0 4YY

22/10/2024

01777633-0009